Praise for
ANIMAL INTERNET

"Bold and fascinating ... proposing that the Internet—and other digital technology—offers an opportunity to rediscover our animals as more than abstracted images but as autonomous individuals with inherent value. A truly thought-provoking book for animal lovers and technology enthusiasts alike."

— *Kirkus Reviews*

"This surprising book offers a great shout-out to the next phase in our relationship with non-human beings: our brand-newly emerging recognition that they, too, are individuals, leading individual lives."

— Carl Safina, author of
Beyond Words: What Animals Think and Feel, and *Song for the Blue Ocean*

"At last, a convincing explanation for why waldrapps are on Twitter and quolls on Facebook. In beautiful, philosophical prose, Alexander Pschera even explains why cats rule the Internet. The first book that brings nature and technology together with animals as individuals and streams of big data alike."

— David Rothenberg, author of
Bug Music and *Survival of the Beautiful*

"A most important book. This excellent work could be a strong catalyst for people to become re-enchanted with all sorts of mysterious and fascinating animals. By shrinking the world it will bring humans and other animals together in a multitude of ways that only a few years ago were unimaginable."

"Humanized pets, industrialized meat, endless sad extinctions: Must our animal future be so bleak? Not according to Alexander Pschera, who envisions humans and wild animals interacting on matters like climate change and conservation through electronic tracking. A fascinating account full of novel and unexpected examples."

"An original book that goes against the trend to stubbornly keep nature and technology divided from one another."

"*Animal Internet* is one of the most interesting books that I've read in recent years."

"What Pschera describes sounds futuristic but it's already widespread reality ... Pschera's book is not just popular science: he describes not only the status quo, but also thinks about an ongoing transformation.

ANIMAL INTERNET

NATURE AND THE DIGITAL REVOLUTION

ALEXANDER PSCHERA

Foreword by
DR. MARTIN WIKELSKI

Translated by
ELISABETH LAUFFER

NEW VESSEL PRESS
NEW YORK

ANIMAL INTERNET

 New Vessel Press

www.newvesselpress.com

First published in German in 2014 as *Das Internet der Tiere*
Copyright © 2014 MSB Matthes & Seitz Berlin Verlag
Translation Copyright © 2016 Elisabeth Lauffer
Foreword Copyright © 2014 Martin Wikelski

Library of Congress Cataloging-in-Publication Data
Pschera, Alexander
[Das Internet der Tiere. English]
The Animal Internet/ Alexander Pschera; translation by Elisabeth Lauffer.
p. cm.
ISBN 978-1-939931-33-7
Library of Congress Control Number 2015950391
I. Germany -- Nonfiction

For D., C., M., and P. in gratitude

Now that they have gone
it is their endurance we miss.
Unlike the tree
the river or the cloud
the animals had eyes
and in their glance was permanence.

John Berger

TABLE OF CONTENTS

ANIMAL INTERNET

FOREWORD

Every few decades, massive societal change occurs, sometimes without our being able to realize or even perceive it at
the time. Walls or dictatorships fall, without anyone having
seen it coming. *Animal Internet* brings this very kind of fur-
tive, yet tremendous, global development to light—involving a total overhaul of our human understanding of nature.
While nature has been conceptualized for centuries as separate from technology, the synthesis of nature and technology
is currently fully underway. In the future, animals' intelligent sensory systems may provide us humans with a trove
of information about life on planet Earth that we need to
survive. Humanity as a whole is thus being provided with
a Seeing-Eye dog that will finally allow the nineteenth-cen-
tury Prussian explorer and philosopher Alexander von Humboldt's ideal of an understanding of nature to be realized or,
as he put it, "to understand the world as a whole through the
interplay of its constituent parts."

Animal Internet: this is the shared, intelligent sensory network that has developed over the course of evolution, and that
animals are now using to communicate with humans. This
store of knowledge will become an inextricable part of the
common cultural framework of humankind, right alongside
libraries and museums and the Internet itself. We are in the
process of incorporating a new dimension of knowledge into
our lives—change like this happens once every few hundred

years and is bound to cause unforeseeable upheaval. Wild animals will become our companions, like untamed and primal versions of our pets that we love and communicate with. The difference is that these wild animals will also warn us, and those humans living in the farthest flung places on Earth, of natural disasters; they will predict the climate; and they will measure chemical levels in the air, water, and soil for us.

Thus we humans will soon be in the agreeable position of understanding the world better through animal behavior, a practice shared by all advanced civilizations of the past. The fundamental improvement is that we will no longer observe only local animal behavior, like the Incas and the bristle worm in the coastal regions of Peru. The Incas used the invertebrate inhabitants of intertidal zones to forecast the weather since they started to migrate when water levels or temperatures changed. But in contrast to the pre-Columbian empire, we will essentially have a worldwide network of bristle worms, because there is no better information system than that of animals, informed as it is by the unparalleled diversity of their sixth sense.

Just as our children today cannot understand how anyone could ever have lived without the Internet, in only a few years we will not understand how humankind could have been so simpleminded—and perhaps technologically arrogant—as *not* to use the endlessly brilliant, perfectly evolved knowledge of the animal world. This book describes that fundamental change, which will be as important to our human sense of self as the discovery of life on other planets.

Dr. Martin Wikelski
Director, Max Planck Institute for Ornithology

INTRODUCTION:
WHY TODAY'S LITTLE RED RIDING HOOD
HAS A SMARTPHONE IN HER BASKET

AN OLD STORY IN A NEW LIGHT

Little Red Riding Hood is relieved. She finally has an iPhone now, too. Sure, her mother said she's still a bit young, and her grades need to improve if she wants to keep this privilege, but the pressure from her clique is simply too great. All her friends have one, and one certainly doesn't want the girl to become an outsider.

Little Red Riding Hood's mother is a single parent, and she's at work all day. She figures it's not such a bad thing for her to know where her child is. Especially since Grandmother moved to that lonely house on the outskirts of town, where Little Red Riding Hood goes most days after school for some adult supervision while she does her homework. The path she has to take goes through some woods that Mom isn't crazy about. She likes that she can get in touch with Little Red Riding Hood, and that her daughter will send her a text every now and then. Mom is also constantly reminding her to take out her headphones when she's in the woods, in order to hear what's going on around her. You never know who might be hanging around there.

But Little Red Riding Hood isn't afraid of the woods, let alone the animals that live in them. She loves the doe, the stags, the fox, and the hare. Every day she discovers something new—usually off the path—which is why she also usually gets to Grandmother's late. But that's not a problem, because when she tells Grandmother about her new discoveries, the old woman beams with joy. For Little Red Riding Hood, animal tracks and birdcalls are messages from friends. Even the wistful "dieu-dieu-dieu" of the bullfinch quickens her heart, and when she sets out toward home at dusk, the intriguing "who-who-who are you?" of the tawny owl doesn't sound like a threat, but an invitation. It is the call of nature, and Little Red Riding Hood is more than happy to listen.

What her mother doesn't know is that this is the very reason she wanted the iPhone. She is as indifferent to texting with her friends as she is to watching those pointless music videos. But all the nature apps out there have given her an entirely new perspective on the woods. Not only can she better identify birdcalls and read animal tracks. Since downloading *Animal Tracker* onto her phone, she now knows that Martha, the vixen living in the den near the clearing in front of Grandmother's house, has four pups. She also knows that the red kite that nests in the spruce at the edge of the clearing struck a different course in returning from its winter home this year. And most important, she knows that a beautiful, big gray wolf has been hanging around the area for days. He comes from a pack along the frontier between Germany and Poland, not far from Little Red Riding Hood's home, and his name is Ferdinand. Ferdinand the Gray, as she secretly calls him, has a GPS tracking device that allows his every move to be followed. What he looks like, what he weighs, which pack he's running with, how many children he has, and everything he's experienced on his forays—Little Red Riding Hood can read up on all of this using her *Animal Tracker* account. A green dot on the map shows Ferdinand's current position. She trembles at the sight of him approaching a main road or the interstate, and always hopes that he'll find a safe route into the nearest woods.

This afternoon, Little Red Riding Hood is atremble once more. Not with fear, but with joy. Because Ferdinand's green dot has appeared on her GPS display and is approaching the red mark that indicates her position. He moves closer and closer. Fingers shaking, she zooms in. Her heart leaps: it can't be more than a few hundred yards separating her from Ferdinand. She carefully looks around. Straight ahead is a meadow, beyond that the woods retreat into darkness. Little Red Riding Hood sets down her backpack and hides behind a mossy tree. She opens the video app on her iPhone and waits. Maybe she'll manage to get some footage. Her breathing grows shallow, and she has to try hard to keep her hand still. Minutes pass, but it feels like hours. Her courage is flagging when she spots a mighty gray shadow moving out of the thicket. A majestic wolf's head appears, frozen in place for a few seconds. Ferdinand the Gray! He assesses the situation in the clearing; it almost looks like he's trying to hear something, himself—hopefully he doesn't sense Little Red Riding Hood. The girl starts recording. Now the animal moves. Slowly, but decidedly, the wolf crosses the meadow. He is moving straight toward Little Red Riding Hood.

The video is running. Two minutes. Two minutes, thirty seconds. Three minutes. Little Red Riding Hood gets a cramp in her leg, but there's no time for the pain. The wolf is now barely twenty yards away. She appreciatively moves the camera along the animal's body—she films his head, back, and bushy tail. The wolf stands still and looks directly into her lens. Has he discovered her? For a fraction of a second, memories of old fairy tales flash through Little Red Riding Hood's mind, images of small, helpless girls deep in the shadowy wood, threatened and devoured by savage wolves. Could there be any truth behind those tales? She thinks of her single mother. "What would she do without me?" For a moment, our uneasy observer considers stopping the video and calling the emergency number.

But it doesn't even occur to Ferdinand the Gray to play into the myth humans have created for him. A warm sunbeam spans the meadow. Steam rises from the grass. The old wolf stretches out his front legs. He yawns with relish and flops onto the grass like a sack of potatoes. Little Red Riding Hood keeps filming and filming. Already five minutes, thirty seconds. What would happen, anyway, if she left her hiding spot and approached the wolf? Would he attack? Or would he simply flee? Suddenly she's tempted to risk the experiment, but reason ultimately prevails. There also isn't time. She's already running over half an hour late, and Grandmother must be waiting for her.

At this very moment, a shrill "beep beep beep" announces the arrival of a text message. Her mother! "where r u … call me right away … im worried … mom" The digital tone shoots through the woodland idyll like a deadly arrow. Ferdinand the Gray jumps up and disappears into the trees like greased lightning. Little Red Riding Hood can just barely follow the green dot on her screen, which is moving faster and farther away from the red one. "Farewell, dear friend!" she whispers. "Take care!"

Her wolf video is six minutes, twenty-four seconds long. No sooner does she arrive at the house than she shows it to her grandmother, who watches eagerly. Together, grandmother and granddaughter watch and rewatch the amazingly sharp images. Little Red Riding Hood can't get enough of recounting the feelings that alternately overcame her, crouched behind that weathered old tree stump: tension, excitement, fear, joy. She suddenly remembers the cramp in her left leg, too. She then uploads the video onto Ferdinand the Gray's Facebook page. The page was created by wolf fans to document the animal's life. The old fellow's got over two thousand digital friends—not bad for such a supposed villain! So far, though, the site hasn't had much more to offer than blurry pictures and tracking updates. Little Red Riding Hood's video is the first long documentation—and the comments left by the wolf community are accordingly enthusiastic. They range from "Fantastic footage, I would have been so scared in your shoes!" to "Forget National Geographic—watch Little Red Riding Hood." Within hours, the video has over five thousand clicks and no end of shares. It finds its way to the World Wildlife Fund and the Natural Resources Defense Council homepages.

Little Red Riding Hood ends up spending the night at Grandmother's house. They drink green smoothies— Grandma is a modern woman, and a vegan, at that. By the time Little Red Riding Hood finally manages to break away from Ferdinand the Gray, it's already pitch-dark. Even the bullfinch has ceased his somber "dieu-dieu-dieu" and tucked his head into his red-gray plumage. Before Little Red Riding Hood pulls the blanket over her own head, she takes out her smartphone one last time, to say goodnight to Ferdinand. And then she sends her mom a text: "hi mom … i made an amazing new friend … his name is ferdinand … tell u about it tmrw … he may be a lot older than me, but he's so beautiful … but ur the best, always have been, always will be … luv u … lil red."

This story could occur, with slight variations perhaps, in much of Europe and North America today. It is only slightly too perfect. The new version of Little Red Riding Hood is no fantasy. It really does exist, this direct connection between human and animal. I have the evidence to prove it: the ultraflat smartphone in my pocket connects me to a flock of glossy blackbirds that spend their summers in Bavaria and their winters in Tuscany: the northern bald ibis, or waldrapp. Software installed on my phone establishes contact with these rare and beautiful birds as they fly in striking V-formation over the Alps for the winter. Many dangers lurk along their voyage over the mountains. This is also why the waldrapp is tagged with a radio transmitter, which allows its handlers to locate and thereby better protect it. The bird's last "encounter" with humans proved considerably less symbiotic. The waldrapp was hunted so aggressively in past centuries that it nearly went extinct, with very few remaining in the wild. Now a reintroduction program is under way, using technology in an effort to bring the bird back home to Germany. The positioning data for individual birds are transmitted via satellite to a database, and from there to a Facebook page, where the information is presented in text, image, and video formats. Facebook is thus transformed into an animal blog. This is a space where the birds—and not just some anonymous representatives of the species, either, but the waldrapp ibises Balthazar and Remus, Tara and Pepe—can share the literal ups and downs of their lives. Using digital technology today, I can get as close to the waldrapp as Little Red Riding Hood got to the wolf, or as the prehistoric hunter had to get to his prey, in order to kill it. The Internet is anything but virtual and abstract here; it is a hyperreal, sensory experience. I can see where the ibises currently are and what they are doing. I can see what company they are keeping. I can see what problems they are combating: whether they have flown into a snowstorm or been blown off course, and how they respond, in an effort to get out of trouble.

A chapter in the history of human-animal relationships has thus come full circle. The hunter of the Stone Age had to come within a spear's throw of the mammoth. He heard the animal's breath and smelled its excretions. Later, the development of better weapons for hunting marked the end of this existential proximity to the wild. It was stored, however, in humans' cultural memory, in the form of fables, myths, and fairy tales, in which frog princes and cat courtiers, foxes and rabbits came into regular contact with humans and one another. After centuries of alienation, humans can now regain this closeness. It is possible to gain a bird's-eye view of the world, flying—like a postmodern fairy tale character—over the ridges of the Dolomites on the back of a waldrapp. We are getting back to our roots culturally, as well. In the grottos of Altamira, the hunter painted mighty animals in earth tones on stone, images that still bear witness to the deep relationship between humans and animals, the hunter and the hunted. My mammoth is a waldrapp ibis that I, along with many others, worship on a Facebook page. These birders comment on incidents that occur during migration and upload their own photos. The digital Facebook wall and the Caves of Altamira are, as surprising as it may sound, two sides of the same coin. Both fulfill the same role in the systems of their respective civilizations. They visualize the awareness we have of nature, thereby strengthening a mutually shared value system.

We are standing on the threshold of a new era of interaction with and awareness of nature, because the waldrapp is no anomaly. There are many further examples of the digital interconnectedness of humans and animals. In the Cascade Mountains of Oregon or on Arizona's Kaibub Plateau, Little Red Riding Hood's encounter with the wolf could become an everyday occurrence—with a better outcome for the animal and without traumatizing the human. Many of the wolves ranging the American countryside, disquieting humans along the way, are wearing GPS collars. Their positioning data are recorded and then transferred to websites, where anyone can follow the animals. In addition, photos captured by hidden cameras provide a vivid take on life in the wolf den. These wolves are no longer anonymous beasts that appear out of nowhere, attack, then vanish in the mist. Instead, they have names and personal histories. They are no longer simply representatives of a species, but rather, real individuals. We can research their biographies and personal preferences as easily as their personalities and social behavior. This ultimately transforms wolves into likable contemporaries—a remarkable career for the creatures, which have for centuries been pegged as humans' archenemies. They have been considered sneaky, sly, ferocious. This new surveillance technology succeeds where generations of ardent eco-advocates and eager ecological educators failed—in freeing the wolf from the snare of misrepresentation and allowing it to turn back into a normal animal. Thanks to digital advances, its daily life has been made transparent and tangible. A new dialogue between humans and nature has arisen. To take this dialogue a step further, it would not be unrealistic to dream of a new language between humans and animals.

But why, some might ask, do we need such a language? Don't we do enough already to research animals' habitats and ensure their survival? Doesn't the conservation debate already get considerable attention in politics, society, and the media? Isn't it even safe to say there's an overabundance of green issues today? It may seem that way subjectively, but in fact, despite a host of international programs, species extinction worldwide is not only advancing unchecked, it is rapidly gaining speed. Human and animal destinies are diverging, humans sitting on the Raft of Medusa, animals floating away on Noah's Ark. The separation has become fundamental. Not only is the basis of many wild animals' existence under threat, but these animals no longer play a role in everyday life, either, having disappeared from regular human routine in Europe and North America over the course of the last two hundred years. In many parts of Asia and Africa, where industrialization has not yet destroyed all native life forms, animals—both wild and working—still serve as companions to humans, sharing in their daily lives, while in the European cultural realm, they were forever supplanted by machines and technical apparatuses from the start of the twentieth century. The key factor, however, is that since then, humans and animals have occupied not only different *living* spaces, but different spaces of *being*— spaces that were formerly one and the same.

But the connection to nature, to that which grows, is inherent to humans, who are and always will be biophilic creatures. Humans cannot live without animals and plants. Hence the green compensation that has followed the years of nature deprivation and loss. Over the last two hundred years, the real animals have been replaced by likenesses. The process is dialectical: the further we distance ourselves from nature, the more we produce, reproduce, and disseminate images of animals—all without moving a single step closer to nature in the process. Postmodern awareness of nature simulates green structures and represents animals merely by pictures, or pictures of pictures, or links to pictures. But these representations do not replace the animals; instead, they simply fake their presence. This awareness has no way of accessing the tangible animal existence on display in the Caves of Altamira.

Nevertheless: overcivilized humans still hear the call of nature. Something profoundly authentic awakens in them when they drive out to the country or read about humans and animals in fairy tales and myths. Even the postmodern awareness of images appears to have an emotional center that responds to the call of the wild. The first questions to emerge: Why is this the case? How can that be? Next question: How can humans respond to this call? Because personal willingness is not enough to break free of the "idea of the idea of nature," the cold prison cell in which the postmodern citizen has been trapped.

Technology, which first brought about and subsequently heightened humans' alienation from nature, is now part of the solution. It is the missing link to reestablishing the connection to the animal kingdom. The Animal Internet has the potential to revive the human-animal relationship, thereby reinventing nature, as it were.

Despite hopes for a newfound closeness, it is worth considering that the shape of nature will face radical change in achieving this relationship. The chaos of wilderness will morph into a networked room. The thicket will be digitally cleared. Nature, defined as the untouched in the Western imagination, will not only be touched, but penetrated. If the postmodern natural world served as the temple of that ersatz religion called ecology, then the natural world remaining after this era will be a desecrated, denuded one. In becoming this transparent, however, nature loses a central quality: nature that is fully transparent is no longer autonomous. This iteration of nature no longer draws its authority from its independence vis-à-vis human society, but rather from its ties to the very same. The Animal Internet is turning nature into a system controlled, if not created, by humans.

This is a watershed in the consciousness of civilization's weary ranks. Nature's increasing illumination and transparency signifies the destruction of a space that has always served as a subjective escape route for the senses. Because in the last few decades, "back to nature" was the last possible antithesis—both politically and individually speaking—to the unrelenting rhythm of the economy and the working world. "Nature" was the last possible stand against the daily grind, perhaps even against what used to be known as "destiny." For many, nature has even replaced God. This nature is now subject to the dictates of the all-consuming Internet. It is being materialized, no longer serving as an idea that can be readily used in opposition to the societal status quo.

The ubiquitous debate about transparency in data gains a significant aspect in connection to the Animal Internet. Today nearly fifty thousand wild, largely migratory animals are equipped with GPS units that transmit a constant stream of wide-ranging data. New animals are hooked up to the circuit every day. Huge amounts of animal data emerge—big animal data, so to speak. Till now, the big data debate has revolved around the questions of how much transparency we want and how people can protect themselves against monitoring by corporations and government entities. This discussion is now beginning with regard to the animal kingdom. Animals are now joining humans in the glass house. Bit by bit, a transparent natural world will emerge—animals are already becoming more easily tracked, cameras with self-timers send pictures from the farthest corners of the rainforest to users' phones; Facebook pages and blogs provide information on animals' whereabouts; smartphone apps reveal the location of highly endangered species like mountain gorillas or orangutans. This revolutionary form of interaction with animals will yield a new, as yet unknown and extremely sensitive store of data available to anyone who owns a computer or smartphone. With this knowledge, humans will push their way deep into the lives of animals; they will break open their "personal space." Against the backdrop of the data debate about the omnipotence of Internet companies and people's vanishing freedoms, the prospect of a *transparent nature* seems like an expansion of the digital war zone. Do we need data protection for animals? The question arises in a genuine debate about the new friendship between humans and animals.

Furthermore, this is necessary because transparent nature introduces a new way of ecological thinking that breaks with conventional ideas and practices of nature conservancy. Transparent nature is no longer a habitat in isolation; rather, it is a green habitat embedded in a gray civilization, the two connected by digital paths, bridges, and tunnels. The central idea of transparent nature is the direct, albeit technologically driven, contact that is again possible between humans and animals. The precondition for this is humans' ability to move freely in the natural world. Habitats and nature preserves, those instruments of classic ecology, function differently. Their objective is not to encourage contact, but to prohibit it. They abide by the principle of systematically barring human access to nature, as opposed to seeing it embedded in their everyday life. Habitats try to save animals from humans. This is reasonable, of course, but it also results in the impassable, ever deepening rift between humankind and nature. Humans are experiencing an estrangement from their natural surroundings, especially from animals. This romantic ecology of exclusion, as one might describe the philosophy of the habitat, attempts to further cast nature as an autonomous space, treating it as if it were exempt from the human-powered dynamic of global development. The central ecological question at the start of the twenty-first century, however, is this: How can humans save the environment without compromising their own future development? How can nature, with the help of technology, figure into the logic of human progress in such a way that both profit?

Humans can save the environment and animals from impending ruin only once they divorce themselves from the conception of technology and nature, civilization and wilderness, as competing or dichotomous spheres. This thinking defines current ecological thinking and concrete conservation practices. Beyond that, however, we must distance ourselves from the myth of an unspoiled, impenetrable natural world that answers only to itself, and acquaint ourselves with the idea of transparency in nature, as fitting to the age of humans. Humans are (inter)connected to animals in a new way in this open nature, and far from being punished for entering this world, they are rewarded. Technology is no longer the eternal adversary of nature, enemy of the righteous and engine of destruction; instead, it has emerged as an ideal, adaptable interface between humans and their natural surroundings.

This book proposes that such radical rethinking is the only way to save the animals. Not because this would automatically result in improved living conditions for them, but because it would teach humans to see animals anew, sensitizing them to their fate. For humans will want to save only what they know. Ecology needs to be restored to its essence: to the tangible relationship between humans and wild animals, which will serve as the foundation for whatever follows. Only a new form of communication can reconnect humans and animals. People today are more focused on developments that directly improve their own lives than on, say, a conservation program for poison dart frogs in Costa Rica. This book suggests a different course. It tells the story of that disruptive moment in technology, when the investment banker in New York or Frankfurt forgets about his new Porsche for just a moment and becomes online friends with a frog.

WHY WE ARE NOW NOTHING MORE THAN BEAUTIFUL SOULS

IN THE LABYRINTH OF A POSTMODERN AWARENESS OF NATURE

At first it was just a knee-high shadow that darted past me, disappearing into the underbrush on the other side of the narrow, winding gravel path. The animal was there one moment, gone the next. But the shadow lingered: it ended up becoming my first animal Facebook friend.

The aviary at Tierpark Hellabrunn, the Munich zoo, is accessible to zoo visitors and consists of an impressively large wire enclosure. Walking through it, one has the feeling of moving freely through the wild. The bird cage is inhabited by colorful ducks, ibises, a majestic black stork, and that scampering shadow that—this much I was able to discern—had a wrinkly red head and a crescent-shaped beak. It certainly wasn't pretty, but the bird was distinctive. But what kind of animal was it? An ibis? A spoonbill?

A sign in the aviary provided the answer. It turned out that the shadow was a waldrapp, a wading bird that, following its almost total extinction in Europe, is now being bred in captivity and released into the wild. The bird won me over instantly with the name "waldrapp." It sounded friendly, onomatopoeic, warm. It reminded me of early Germanic history, evoking deep, dark pine forests. Vignettes from the old folk songs and poems collected in the anthology known as *The Boy's Magic Horn* swirled through my mind. I could imagine the waldrapp on a punched copper frontispiece adorning a yellowed tome of romantic verse. The bird's Latin name, on the other hand, had rather a jarring effect. It snarled at me in a gravelly voice: *Geronticus eremita*. "Gerontic eremite"—it seemed to suggest organized bus tours for senior citizens swathed in blankets. Without further ado, I decided to leave the astonishing creature alone and continue my exploration of other parts of the zoo, rather than pursuing a more extensive interaction with this peculiar bird. After all, Hellabrunn has many interesting—not to mention, actually beautiful—animals that do not take cover at the sight of visitors and that furthermore do not remind you of your own future decline. Farewell, then, waldrapp.

For now. I couldn't quite get the weird bird out of my head. Something about the way it sounded … It was the poetic name that stood out most in my memory. I knew that I had encountered this animal before, but where? After a long search, I found the source of my memory. It was Anita Albus's *On Rare Birds*, a book of portraits of extinct and endangered birds, presented in eloquent language and exquisite images. There are chapters on the kingfisher, the northern hawk owl, the corncrake—and, as it so happens, one on the waldrapp. Most important, there is a picture of a small waldrapp colony that immediately sparked a feeling of recognition during that brief encounter at the zoo.

Anita Albus's portrait of the waldrapp is not a naturalistic illustration. It looks almost like a medieval panel. It is suggestive of paintings by Lucas Cranach or Albrecht Altdorfer and could easily hang in the Alte Pinakothek art museum in Munich: overlooking an idyllic riverbank that dissolves into the blue of a distant horizon, a brilliant member of *Geronticus eremita* sits enthroned above a city dotted with towers, a seemingly enchanted cliff face serving as its backdrop. The bird's greenish-black plumage takes on the landscape's color palette, a symbol of the unity of animal and environment. No one would dream of using so prosaic a term as "habitat" at the sight of this idealized landscape. The natural world depicted here knows no habitats, no delineated preserves, because it all still belongs to the animals. The animals are at the heart of this world: the waldrapp's yellow eye glints at the center of the picture. Its gaze speaks of wisdom and perspective, dignity and supremacy. This eye is the gravitational center of the preindustrial cosmos, before it all spun out of control.

So there it was, my Hellabrunn shadow, but rather than some timid phantom of a bird, it was a fine ornithological specimen, captured on thick canvas in choice, handmade pigments and printed in a handsome volume. There was just one thing I could not reconcile: of these two animals, the one had nothing to do with the other. Despite concerted attempts at association, the real bird defied connection to the panel, as the aura of the two creatures was simply so different. There was even less connection between the picture and the waldrapp ibises living in the wild, which are currently being reintroduced in Europe. These animals have forgotten how to migrate and must be guided over the Alps by an ultralight aircraft. Many are lost in the process. Others are plucked from the skies by trigger-happy Italian hunters. The life of the modern-day waldrapp is anything but idyllic. Its existence is a scientifically monitored project with many dangers and setbacks. Anita Albus's fantastical illustration has nothing to do with this reality. The plumage looks deceptively real. The iridescent play of colors enchants viewers, who can hardly tear themselves away from the old-world precision of the painting. This precision does not, however, translate into a palpable presence. The exactitude of the brushstrokes does not call the bird to mind, allowing it to come to life and demand our involvement. Rather the opposite occurs. The bird is relegated to a timeless, distant place. It resembles a mythical creature, an animal drawn from poetry and fairy tales.

The expository text does its part in emphasizing the bird's legendary character. The waldrapp has, it says there, gone by many poetic names, which suggests that difficulties arose in trying to classify it zoologically. It was called *Schopfibis* or *Mähnenibis* (crested ibis), *Klausrapp* or *Klausrabe* ("hermit raven"), *Steinrapp* ("stone raven"), and finally, *Waldhopf* (forest wood hoopoe). The Zurich-based ornithologist Conrad Gesner, whom Albus references extensively, called it simply *Schleck*, a culinary delicacy, which helps explain why the waldrapp stood no chance at surviving the seventeenth century: its meat tasted "so exquisitely sweet." The unsightly waldrapp could hardly have served as a grand trophy. The description Gesner gives, as one can read in Anita Albus's work, supports this suspicion. He describes the bird as "quite black if you look at it from a distance, but if you look at it close by, especially in the sun, you will consider it mixed with green. Its feet are also somewhat like a hen's, but longer and the toes split. The tail is not long. It has a crest on its head pointing backwards ... The bill is reddish, long, and suited to poke with it into the ground, and into the fissures and holes of walls, trees, and rocks, to extract the worms and beetles which hide themselves in such places." By the nineteenth century, the memory of the waldrapp had already grown so faint that people assumed it was a mythical creature. Anita Albus captures this memory of a memory in her illustration, just as her book is a collection of *memories of memories* that people have of rare birds.

Then all of a sudden, I understood why the two images—that of the darting shadow and the bird painted in the style of the Old Masters—were uncoupled, indeed, why they contradicted each other. There are two versions of nature: the first is nature as we picture it. The second is nature as it truly exists. This second nature—the real, raw, wild one—has left us. The wilderness, which we have not seen for so long, let alone visited, exists in our minds as a narrated, painted, filmed, and artistically photographed reality. These artifacts range from the works of Jack London to filmmaker Heinz Sielmann's *Expeditions Into the Animal Kingdom* that defined my youth and that of many of my European contemporaries, from IMAX nature documentaries in high definition to Anita Albus. These are fragments of a massive arsenal of metaphors protecting the memory of nature and shaped by two fundamental forces: sentimentality and nostalgia. Anita Albus's book is an extreme manifestation of this nostalgic-sentimental look back at nature that is less about the desire to return to the wild, wading through the muck and dirtying one's hands, than it is about the urge to transform nature entirely into culture. The goal is artistic perfection, in which nature "as nature" is extinguished in a gesture of aesthetic elevation and preserved thereafter as a symbol. It is a desperate attempt to grasp at what is slipping through our fingers, but it is also the final triumph of man over nature. Although the painted reconstruction may have started as a rescue attempt, its gesture of romantic overwriting reveals it as a form of mummification. This attempt represents the pinnacle of the romantic misconception of nature as a form of beauty, rather than as a form of reality, of objective actuality. The waldrapp—that bald, shy bird—becomes a hyperreal aesthetic vision in Albus's representation, a fantasy image that shows how "it" may once have lived, and how sadly "it" will never be again. The false symmetries of culture have desensitized us to the real ambiguities and ugliness of nature. And even when we do try to escape the constraints of beauty—as amateur environmentalists, or ornithologists, for example, who organize group trips, study the literature, and travel abroad in the name of nature—we are still driven by a cultural motivation: we want to impose order on our observations, collect bird species, have unique experiences, breathe life back into our green self that's been deformed by society.

Pictures of the waldrapp thus supersede the reality of the waldrapp. Pictures of animals push the actual animals out of the way. This is a defining characteristic of the postmodern awareness of nature. My afternoon in the Munich aviary was a pictorial flight of fancy, and not a walk through the actual natural world. It showed me that we do not, indeed that we *cannot* truly see animals anymore, let alone touch them. We content ourselves, instead, with remembering them.

DO NOT TOUCH!
THE REPERCUSSIONS OF LOST SENSORY EXPERIENCE

Our loss of the animal world manifests itself as the loss of our elementary sensory capacity. Beyond the realm of house pets, which have become active participants in our society at this point, there's no remaining sensory connection where humans and animals can meet. Fundamental physical contact—petting, cuddling, brushing, and milking, but also slaughtering, gutting, cutting—is essential to this relationship. Care and carnage go together. They're two sides of the same coin. Direct, bloody encounters with the animal world—like slaughtering a rabbit, gutting a carp, or plucking a chicken, which older generations still remember—are viewed today as barbaric acts that few could bring themselves to do, even if they're not vegetarians. With regard to animals, we are no more than what Hegel called "beautiful souls" who dream of peaceful coexistence while eating factory-farmed chicken. The murder must occur outside our range of perception. Contemporary cooking plays its part with its abstract contortions, transforming a simple piece of raw meat into a dish served in fine restaurants, arranged on the plate like a Kandinsky painting. Meat reaches us as a biomass that is uniform in color and artistically presented, and that's not allowed to reflect its animal origins, but must instead obey the law of aesthetics. A cultural transformation in the way living creatures are perceived can clearly be seen here. Many foreigners visiting France, for example, may be unnerved by the displays in French butcher shops, where scaly skin may be left on the chicken legs, patches of fur on the rabbits. These last remaining signs of animal life come as visual shocks, although they really should indicate the quality sourcing and naturalness of the meat. But these fragments of the wild that find their way into civilization and consumer society deeply disturb our sensibilities when we set eyes on them.

Why? Because over the last fifty years, an unfortunate change has occurred in the way we raise children that has systematically distanced us from animals. Many of the great natural scientists and nature writers have childhood memories of entire afternoons spent catching animals and collecting flowers. Insect collections and herbariums used to be commonplace features in any childhood room and might serve as the starting point for an expedition into the natural world. Truth be told, if one wants to get close to nature, one must touch it. This is the secret of gardening. And it certainly applies to animals. One who has never torn off the tail of a lizard knows nothing of the reptile's slippery agility. One who has never sunk thigh-high into a pond's stinking mire while trying to catch a frog knows nothing of these amphibians' leg power. One who has never tried to catch a bat that mistakenly flew into the house on a hot summer night and took cover under the bed, knows nothing of how viciously these little mammals can hiss, and what pointy teeth they have. One who has never chased a butterfly hither and thither knows nothing of how unpredictable these insects' flying maneuvers can be. The secret to animals is revealed only through a sensory, a physical act. To grasp is to understand. A relationship can form only once one has touched an animal's skin or fur, felt its teeth, smelled its warm excrement in the palm of one's hand, or destroyed its delicately pigmented wings in a thoughtless move. Only then does the animal leave its imprint on the senses, and thus on the person's life story, becoming part of that story. And only then is that person prepared to do something for this animal, because now the "it" has become a "you." It has become a friend.

This reciprocal exchange with wild animals has become impossible in today's world. Wild animals are foreign creatures that have been recast in a symbolic role, recognizable at best from documentary films or zoos. They live "out there" in the natural world, tolerated by civilization, inventoried by biologists. For most people, concrete contact with wild animals is taboo—unless they submit themselves to the overregulated hunting and fishing system, which has itself now mutated into a type of romantic, overblown pest control. Nature doesn't belong to us anymore. This is the aim of nature conservancy, which has put strict demarcations in place to protect biodiversity. No one can say who nature belongs to, though—not even those working to protect it. This is because it has morphed into "nature as such," an abstract construct which society keeps referring to, but that simultaneously demands our utmost consideration. But why should humans care about something that has nothing to do with them, that they are not allowed to touch? It comes as no surprise, then, that for most people, the natural world—even right outside their back door—is something alien that has no bearing on their life. It's an idea, an image, a cultural phenomenon, a supposition.

Not so very long ago, this was fundamentally different; smell and touch constituted an awareness of nature that is as endangered today as many animal species. Many authors, in whose works nature plays an important part, have described how their sensory contact with nature—which was not infrequently quite dangerous—provided them with a deep understanding of animals. It is worth reading these stories closely, because they illustrate the distance that separates us today from existential contact with nature. In his memoir, *Green Branches*, a thick compendium of sensory experiences with the natural and animal worlds, Friedrich Georg Jünger, the younger brother of the German author, aesthete and entomologist Ernst Jünger, describes how, as a child, he was able to locate the vipers living in the stone quarry by their "curiously pungent smell" alone. The nose, the olfactory organ, establishes the first, most direct connection to nature. Like hunting dogs, humans pick up traces of animal scents with their noses. This form of approach originated in the distant past of human development. Together with his brother, Ernst, Friedrich Georg Jünger collected all manner of beetles by "knocking, sifting, chomping, prizing off" sections of tree bark. Every form of contact—as reflected in a grand, almost lyrically onomatopoeic repertoire of action words—is employed to get at the animals. The boys' driving curiosity is not slowed by unknown quantities: "ground and bark mushrooms, rotten fruits, carcasses and excrement were examined; carrion, cheese rinds, and other bait laid out." The interaction with nature becomes a full-body affair: "In order to be totally unhampered, we undressed and hid our clothes in an alder bush and wandered naked half the day through the marshy meadows and reed thickets that surrounded the water in a wide, green band. To repel the mosquitos, gnats, and horseflies, we smeared the thick, black sludge all over our bodies ... Then we hurried with quick steps over the turf of the floating meadow that undulated under our weight, like a calm lake."

In the foreword of his wonderful book on cranes, zoologist Josef H. Reichholf also tells the story of how he, as a ten-year-old child, stole a young jackdaw from its nest in a church tower and raised it at home. The hunting excursion proceeded as follows: "The jackdaws had been nesting in this tower for centuries. They built nests in the rafters and added layer after layer every year, till one of these towering nests grew too tall and toppled. Fragments of these nests that were full of fecal remains, dust, and the mummies of baby birds that had never taken flight, landed far below on the upper platform ... Pretty filthy from all the stuff that had rained down on me—because I inevitably bumped into old nests—but with a screeching baby jackdaw as my haul, tucked away under my shirt, I came back down and slunk away from the church like a thief."

The humans who feel drawn to nature take it into their possession. They instinctively grasp at it. The beetles end up in a display case, the jackdaw flutters through the aviary, a terrarium is assembled for the slowworm, which may not last longer than three days in captivity if not properly habituated. Living creatures are killed, collected, pinned, or caged. Destruction is not the aim of those who grasp at nature in this way; in fact, quite the opposite applies: they want to join in the natural order as something that belongs there.

I can still clearly remember the first vivarium I had as a

child, in which I kept a sand lizard for two days. The vivarium did not have a lid and sat outside in the yard. One morning I discovered the lizard dead, with a deep hole in its back. The creature at once became a fascinating object of close study. The lizard had been killed by a bird. This event revealed the lizard's vulnerability to me. Danger came from above, it turned out: the reptile's Achilles' heel was the blind spot on the neck where the eyes of the otherwise very quick and observant animal did not reach. The lizard was dead, after only two days, and now a new specimen had to be caught. This behavior wasn't destructive, and it certainly wasn't wiping out any lizard populations, but it did result in bringing a child closer to nature. It sparked an interest that has endured until today. Making nature one's own or appropriating nature, which itself primarily obeys the law of the jungle, is not an act of annihilation, but one that engenders respect. It enables people to study up close what would otherwise remain abstract. Children today can no longer experience this nearness. What are the reasons for this loss of sensory experience?

A central problem stems from the fact that modern science and its teaching systematically undermine the value of the visible. This has led to the devaluation of the perceptible world, as opposed to unseen structures. The ideology of the invisible has especially taken its toll in schools. Biology instruction no longer covers practical zoology, concentrating instead on abstract connections. Instead of species identification, teaching now focuses almost exclusively on molecular biology and genetics. The result is the loss of openness to phenomena, the atrophy of human senses and sensory nature. The tangible world is being degraded, as opposed to the unseeable world of molecules and enzymes, the structures that constitute our world. Over fifty years ago, in her book *The Human Condition*, Hannah Arendt already drew attention to the danger of sensory experience disappearing from the scientific process, the roots of which she traced back to Cartesian thought. In doing so, she analyzed the connection between the loss of fundamental common sense and the consequent effect on society and its political discourse, based now solely on abstract deduction, rather than on visible reality: "This faculty the modern age calls common-sense reasoning; it is the playing of the mind with itself, which comes to pass when the mind is shut off from all reality and 'senses' only itself. The results of this play are compelling 'truths' because the structure of one man's mind is supposed to differ no more from that of another than the shape of his body ... The Cartesian solution of this perplexity was to move the Archimedean point into man himself, to choose as ultimate point of reference the pattern of the human mind itself, which assures itself of reality and certainty within a framework of mathematical formulas which are its own products."

Also contributing to humans' estrangement from their natural surroundings is the myriad of technological aids that permeate daily life, the extensive use of media, the intense allure of the digital world. Nearly every recent study of the relationship between humans and nature shows that the high level of digitization and virtualization in our society is to blame for our diminishing connection to the natural world. The Internet, smartphones, and GPS systems are made out to be the causes of a global acceleration obscuring the rhythm of nature. The flood of stimuli in the digital world has brought about a deadening of the senses. And truly: we are scarcely able to perceive nature or move freely within it anymore. Natural space now serves as little more than a stage for athletic activity. We rely on technological aids to the extent of unlearning how to read nature's signs and orient ourselves according to them. And every hour spent staring at a computer screen is another hour not spent out in the fresh air.

There is a further reason for this estrangement that is by far the most telling. Because even without the distraction, indeed the reeducation brought about by technological devices, it would be very nearly impossible today to explore the natural world inhabited by wild animals. Those places that remain biodiverse, removed, and wild are now mostly inaccessible. The nonurbanized world is demarcated by countless boundaries and governed by restrictions. Nature preserves alternate with land trusts, while between them stretch exhausted fields that cannot support much more growth than dandelions. Hobbies that were once commonplace and established the very basis of a human interest in nature can now be considered criminal: foraging for mushrooms, picking flowers, catching and observing animals, filling butterfly display cases and curating insect collections. This all used to be a self-evident part of a creative way to make nature one's own, a way for humans to immerse themselves in natural space. But now going after a cabbage white butterfly or slowworm can result in a hefty fine for the perpetrator. These days, parents planning a butterfly collection or catching a lizard for their child's terrarium are engaging in illegal activity. But a sense of freedom, or even anarchy, is a fundamental part of discovering nature. For children, nature is the antithesis of the world of school and parents. Animals are ambassadors of a different, freer life. Forming a connection to them amounts to breathing the air of a better world. Contemporary nature conservancy has managed to morally recast this refuge and poison the air of freedom. The Eleventh Commandment reads, "No Touching!" And those who do so despite the warnings suffer a guilty conscience, given the extent of our reeducation. We can all easily assess this in ourselves.

The result of this general prohibition on touching manifests itself in younger generations as a massive deficit in knowledge of biological connections, plants, and animals. It comes as no surprise that the ability to identify wild species is in rapid decline, as many recent studies show. Most people don't even know what kind of animals live in the woods beyond their own backyard. They can no longer identify birdcalls or read animal tracks. Even common native species like buzzards or badgers are no longer recognized. The loss of connectedness to nature has become pathological. There is even a condition called "nature deficit disorder" that is characterized by this very disorientation in natural surroundings. Far from an alarmist horror scenario, Richard Louv's successful book *Last Child in the Woods* depicts the bitter reality. Louv recently concluded, apodictically, "The more high-tech we become, the more nature we need." According to the author, the result of this growing tech dependence is a clinical condition that can be combatted only by systematic greening efforts and regular encounters with nature. Rather than the World Wide Web, Louv recommends what he calls the "Web of Life," as part of a therapy program he has designed and conveniently made available for purchase.

However exaggerated this therapeutic approach may be, Louv is justified in his diagnosis. The hypothesis suggesting humans' fundamental distancing from nature is empirically substantiated. The 2010 annual report of the Children & Nature Network determined that "children's recognition of wild species continues to decline." This can be traced back to the fact that for years, visitor numbers at U.S. national parks have been in steady decline. The numbers have decreased by 7 percent between 1997 and 2010. Another study identifies the clear connection between declining numbers at national parks, on one side, and the increase in use of electronic gadgets, on the other: "This decline, coincident with the rise in electronic entertainment media, may represent a shift in recreational choices with broader implications for the value placed on biodiversity conservation and environmentally responsible behavior."

The statistics are clear: America's children today spend over fifty hours a week with electronic devices and less than one hour outdoors. Just one generation earlier, children spent at least four hours a week out in the fresh air. Children who grow up biased toward electronics will not grow into adults concerned with nature conservation—and it's because they feel ambivalence toward nature or, more dramatically, because at some point, they won't even know what could possibly be meant by "nature."

FORMS OF COMPENSATION IN THE CONTEMPORARY AWARENESS OF NATURE: BIRD-WATCHING, ZOOS, HOUSE PETS

Now it can't be said that we've lost all interest in animals. In fact, one almost gets the sense that human involvement with nature is greater than ever before. Nature observation has become an expensive hobby, zoos are logging record visitor counts, and there have never been as many house pets on Earth as there are today. Upon closer inspection, however, our obsession with animals reveals itself to be an obsession with ourselves. These are forms of compensation that, rather than bridging the rift between humans and animals, only serve to deepen it. This compensation puts nature at a distance and makes it something abstract, something untouchable.

The concept of *noli me tangere* again applies here. Touching is supplanted by viewing. Schoolchildren learn to recite the stubborn mantra of "just look, don't touch." Adults then carry on the "just looking" rule to its consummate end in the newly popular sport of bird-watching. The semiprofessional bird-watcher spends a small fortune on "optics," as they are called; in other words, on the binoculars and spotting scopes. This is a paradoxical investment in an instrument that separates the bird-watchers from nature, keeping them at a distance while giving them the impression—thanks to longer focal lengths and higher resolutions—of getting closer than ever to the animals. Bird-watching illustrates the dilemma at the heart of the postmodern awareness of nature. An imagined closeness is created that actually amounts to the most profound separation.

The bird-watcher is fundamentally solitary in relation to the animals. The solitude is reflected in the birder's activities that, however driven by the hobbyist's zeal, ultimately comprise little more than counting individual animals and collecting glimpsed species. Both activities are forms of a catalogic aesthetic that subordinates nature to the logic of the observer and that focuses more on the observing subject than on nature. Ornithologists uphold an aesthetic biophilia that is long established among the educated and that remains shaped by structural categories dating back to Goethe and the Enlightenment. Nature serves here as an intellectual refuge for the subject in search of self, jaded by civilization, who ultimately finds salvation in the purity of natural experience. The current bird-watching trend serves as a good example of the cultural transformation that has changed what was once an existential tie between humans and animals into a reactive stance of pure observation.

Ornithology is one form of compensation for a modern society that has lost touch with nature. The second form of compensation is the aesthetic reconstruction of animal freedom, as found in zoos and safari parks. More and more money is being poured into modeling deceptively realistic nature dioramas. These reconstructions of freedom entail the principle of revealing and concealing. What happens is that the animals we bought tickets to see at the zoo hide from view inside their papier-mâché caves. Visitors accept this; in fact, we demand it. And why? Because that moment of surprise, of unexpected discovery of an animal that has dared emerge from the safety of its imitation habitat, replicates the very feeling that overcame our ancestors when encountering an animal in the wild. One could almost say that the average person goes to the zoo to *not* see animals, and for the chance to relive that lost feeling of freedom. To *not* see animals at the zoo creates a curious tension of expectation that seemingly makes us revert to the role of hunter-gatherer on the savanna.

Visitors to the zoo regularly encounter ruptures and

jarring scenes that betray the fictional nature of the natural experience created there. The notion of the zoo relies on the theatrical staging of freedom. Any given scene, however, is embedded in the real world, which repeatedly breaks through the illusions it is trying to create. The visitors then realize they are living in a compensatory system, in which they are unable to describe what they see. A personal anecdote may help to illustrate this existential distance from animals that characterizes the zoo experience.

A sunny October afternoon in Munich. The Oktoberfest is over. FC Bayern Munich is already ahead in the soccer standings. The city exudes a self-satisfaction that stands in open defiance of autumn. The trees along the Isar River have changed to red, orange, ochre. Glasses clink in the beer gardens. My family is on a trip to the Hellabrunn Zoo, one of the few outings that both children and parents can happily agree upon, and that has therefore become a regular destination. We stop at the brown bear enclosure, which currently houses a mother bear and her cubs. The kids push their way through the crowd and press their noses against the thick safety glass. "Oohs" and "ahs" fill the air. Look how adorable those little bears are! A shadow suddenly appears—a group of ducks flies in and lands on the small pool that separates the bears from the visitors. A mother and three sweet little ducklings. It all happens very quickly: the mother bear doesn't waste a moment. She plunges into the water, scatters the ducks with a swipe of her paw, and devours the small birds. Feathers are all that remain. Then she trots back to her cubs, which have been basking in the autumnal sun all the while. The visitors gape in horror. The first sobs emanate from the front row. Cries for Mommy. Mothers and fathers think frantically: What do I say to my kid? Honestly, what? Because that can't possibly be the nature we set out to see this morning. I am one of those fathers who could not come up with a very good answer to that question.

It became clear to researchers at the Zoological Society of London that we no longer possess the language to adequately describe what we see in nature, because we have lost the sensory connection to them. The scientists are on a mission to photograph the Siberian tiger. What their cameras have captured, however, is a different natural display altogether: a golden eagle attacking a sika deer many times its size. It's an occurrence very rarely seen. The truly spectacular photos circulate through the media. In the German daily newspaper *Die Welt*, one journalist pens the headline: "Rare, Brutal Natural Display Caught On Camera." The article reads: "A golden eagle attacked an unsuspecting sika deer in southeastern Russia. It dug its talons into the animal until it died." The word choice is revealing. The "brutality" of the event and the aggression—like the "unsuspecting" nature of the deer—are in the eye of the beholder judging the content of the footage. The same justification could be used to describe the "brutal" way in which the sika deer eats the bark off the "unsuspecting" birch. Et cetera, et cetera. The moral reflex embedded in our language shows our distance from what truly happens in nature. A tenor of sentimentality takes precedence, ascribing moral coordinates to essential natural processes and rendering the facts of nature literally indescribable.

Our lack of awareness about this distance can be attributed to the fact that we are now surrounded, on a daily basis, by humanized animals in hitherto unseen numbers. There have never been as many domestic pets on Earth as there are today. There are now over 160 million in the United States, well more than double the number in the 1970s. But these dogs, cats, and birds no longer play the role they once did. House pets used to serve humans. Today's house pets are less animals that serve a distinct purpose, and far more integrated members of the family. They are part of their owners' social networks. Pets have been elevated to morally qualified beings. This is a relatively new development in human history. In the early nineteenth century in France, dead dogs were simply discarded in the Seine. Things began to change only half a century later. Suddenly, Parisian high society began interring its pets in special pet cemeteries and adorning mantelpieces with stuffed dachshund heads in fond remembrance of the deceased. Dogs were outfitted with their own wardrobes, complete with boots, bathrobes, and bathing suits. Animals became social creatures, ersatz humans, as it were—and it is how they are treated to this day. We give them gifts, we give them human names, we dress them up and bury them like humans.

In reality, though, today's house pets are little more than shadows of their ancestors. They are usually sexually isolated and confined to small spaces. They have little to no contact with other members of their species and are given artificial food to eat. In providing people with reminders of the natural world, they have become ossified, living pieces of furniture, props on the set of their owners' life stories. House pets are hereby drawn into a cultural constellation that reduces them to mere toys, to home furnishings, or even to a symbol of a type of love many believe has ceased to exist in humans. The weary warrior of the civilized world finds a therapeutic, fetishistic release in the outcome of this transformation. Today's house pets are no longer worthy counterparts in an existential exchange, but rather artifacts that need to find their place in our feel-good culture. It ultimately amounts to the same, whether they are living or stuffed and sitting on the couch, like so many dogs, cats, and horses that came before, and that were bound to be immortalized, their bodies manipulated into a favorite pose and stuffed with straw.

Bird-watching, zoos, house pets—all three forms of compensation merely give the impression of animal closeness by creating the cultural constellation onto which images of animals and snippets of nature scenes are projected. This is the most visible logic behind the compensation: nature is superseded by pictures of nature. The estrangement from animals is directly palpable in the untold number of animal images surrounding us. The more animals vanish, the more pictures of them appear. Humans try to hang on to the tangible reality slipping away by taking a picture of it. The desire to hang on to things signifies a fear of loss. Cell phone photography casts an unsettling light on our relationship with eternity. The people viewing a cathedral through the screen of their smartphone have lost sight of what this architecture represents. The faster we forget the animals, the more urgent our need to hang on to their image in photographs. This is an act of remembrance, a nod to the animal as man's evolutionary partner.

THE TRUTH BEHIND THE PICTURES

Animal pictures allow us to lose sight of the real disappearance of animals that is currently under way and gaining speed, as recent studies show. It took humans 23 million years to eradicate about 10 to 20 percent of the animal species on the planet. According to the World Wildlife Fund's *Living Planet Index*, the earth has lost half its animals species in the last forty years. In the meantime, human destructiveness has intensified to the point where, if we continue at our present pace, we will have wiped out a further 20 percent of species in the next thirty years—plants, insects, spiders, amphibians, reptiles, birds, mammals. One does not need to be a math genius to comprehend the catastrophic ramifications of such a rapid loss of biodiversity. Half of all known species now face extinction. The factors responsible for this are the destruction of habitat, pollution, human exploitation of natural resources, climate change, invasive species, epidemics, wars, and geopolitical transformations.

Current studies show that half of the North American bird population is endangered. This includes both urban bird species like the Baltimore oriole and rufous hummingbird, and those found in the wilderness, like the common loon and bald eagle. A report published by the Audubon Society states that 314 species would face a massive decline in numbers, should global warming further worsen. This is cited as a key factor in extinction rates, because changes in temperature force animals from their native habitats. The bald eagle was once the poster child for American conservation efforts. It is now poised to lose up to 75 percent of its population by 2080. Between 1900 and 2010, destruction of freshwater habitat in the United States advanced at a rate 877 times faster than in prehistoric times. Most recent estimates assume that the rate of extinction will have doubled by 2050. Between 1898 and 2006, the United States lost thirty-nine species and eighteen subspecies. Researchers now predict that a further fifty-three to eighty-six species will have gone extinct by 2050. The causes all stem from human interference with wildlife habitat. Freshwater fish are a good indicator of general extinction patterns, because wetlands are home to a broad diversity of species. Fish extinction is just the tip of the iceberg. Mussels and snails are dying out even more quickly.

The United States is typical of the current global trends of animal extinction. In order to get a feel for these fatal developments threatening the animal world, Edward O. Wilson, who dedicated his entire life to the study of ants, performed a quick but telling calculation: the current understanding is that over half the life forms on Earth live in the tropical rainforest. Every year, 1.8 percent of the rainforest is destroyed, which translates into a loss of 0.5 percent of the world's species. If the rainforest is home to 10 million organisms, which is a conservative estimate, then fifty thousand species are lost per year, 137 per day, six per hour. And this estimate is not even the worst-case scenario, because this calculation of extinction rates is limited to the connection between surface area and species. Further negative influences, pollution, disruption, and the introduction of foreign species are not factored into the projection. Prospects are similarly desolate in other diverse habitats, such as coral reefs, river systems, lakes, and wetlands. There is, of course, enough wilderness left on Earth that new animal species are still being discovered—and not just bacteria and inebriates, but mammals, too, like the Andean olinguito (*Bassaricyon neblina*) and the Flores giant rat of Indonesia, which typically live in remote areas. But these are rare exceptions to a sad rule. At the same moment researchers gleefully add the odd new species to the register of planetary inhabitants, the overall development of their own species results in whole pages of that register being wiped out. Should the current trend continue unhindered, by 2030, habitats for apes will have been destroyed by 90 percent in Africa, 99 percent in Asia. The earth is becoming a planet of dying apes.

When faced with the logic behind this fall to ruin, it is hard to remain optimistic. This is especially the case when one considers the ineffectiveness of the animal protection initiatives that heads of state so often sign in order to gain popular appeal. It has reached the point where one could justifiably adopt the defeatist stance of the "nature pessimist," who considers the earth's fate to be sealed, since the course of extinction is no longer stoppable. For example, there are far too few tigers and apes left for these animals to muster the strength to survive and reproduce, should their current situation worsen further. Ground cover has grown too thin. We may be the last or second-to-last generation to wander the earth with wild, free-ranging mountain gorillas or orangutans. We may need to accept this logic of extinction as the flipside of an increasingly technological world in which the Western standard of living is the guiding light that everyone wants to follow—and should.

Despite numerous sweeping measures—one need look no further than the 1992 Earth Summit in Rio de Janeiro, or the Convention on Biological Diversity in 2002—the facts indicate that not only is the rate of extinction on the rise, the earth's general health is deteriorating. It is clear today that every environmental problem discussed in Rio in 1992 has worsened significantly in the intervening twenty-five years, rather than improving. Between 2000 and 2011, worldwide carbon dioxide emissions raced unimpeded from 24.9 to 34 billion tons, thereby achieving a record high. The world is currently losing thirteen million hectares of forest yearly. Desertification is also advancing unchecked.

The naïveté required to envision going "back to nature" in the face of this dynamic of destruction is astounding. The natural world to which humans like to think they could return, in turning their backs on civilization, no longer exists. In fact, it never has. The *state* of nature is not some static construct; rather it is caught in a system of perpetual change, continually reshaping itself to accommodate the conditions affecting nature. The image of a realm free of all human influence has been desirable since the days of Rousseau, because it promises liberation from the shackles of present concerns and suggests a return to innocence. Today's way of thinking about natural habitats and the praise of wilderness without humans is still informed by this push "back to nature." Yet the concept remains utopian, an idealized cultural product created by humans—and later, we will see how our conception of ecology and environmental protection is defined by this culturally derived notion of a natural world at odds with civilization, and how we are missing the target of effective conservation as a result.

The question needs rephrasing. Not, "Can we return to nature?" but rather, "Can we view nature differently?" and, "What will happen to us, the viewers, once we have seen nature as never before?" There is no way back to nature. What may well be possible, however, is the emergence of a new *image* of nature—an image that is concrete and stimulates the senses, that breaks through the abstraction and doesn't just give the illusion that everything is fine and under control, not unlike a parakeet chirping happily in its cage on the windowsill. This new image would then be used in establishing a new paradigm for encounters with nature. And it is here that the Animal Internet begins.

WHY WE KNOW WHETHER A SWALLOW IS FRIGHTENED IN A STORM

WHAT REALLY HAPPENS ON THE ANIMAL INTERNET

We currently know little more than one percent of the way animals live in the wild. That is to say, little more than nothing at all. We don't know how little sea turtles behave just after they have slipped out of their eggs on the beach. We don't know how a young cuckoo finds out where to go when autumn breaks. This is why it is also so hard to protect animals. Fundamental connections between the life of animals and their surroundings remain unknown. We know far too little about the lives of many endangered animals to help them effectively. We don't even know if some species still exist. New species appear, or reappear, every year. Take, for instance, the Spotted-tail Quoll (*Dasyurus maculatus*), a carnivorous marsupial that was recently caught on remote digital camera in Australia's Grampians National Park, after being presumed extinct for 141 years. The last of these animals was supposedly killed in 1872. They were considered a real pest then. The presence of this species, however, provides a wealth of information. It is a sign of a stable ecosystem, because, as a nocturnal carnivore, the Spotted-tail Quoll occupies a spot at the upper end of the food chain, therefore inviting comparison to the Tasmanian devil. If this animal has survived this long, then the same must be true of the species on which it preys. Generally speaking, however, humans' prior knowledge of most animals is so minimal that it is impossible to deduce any further understanding from it. It does not provide any reliable empirical foundation upon which actionable strategies could be built. Every year, billions of birds and bats fly thousands of miles from their breeding grounds to their winter homes. What actually happens during these migrations, however, remains a mystery. What we know is that mortality rates are very high during migration. But what we don't know is when and where highly mobile animals die. In many instances of endangered species, we cannot answer the question of what exactly we need to protect in order to save them: Is it food options? Water quality? Botanical diversity? What prevents us from creating a telling picture of nature and formulating effective rules for humans' behavior toward it, is the lack of hard, empirical data and concrete information: What animals currently exist? How do they move around the planet? What do they do underground or at night? Whom do they eat, and who eats them?

The data gathered from the Animal Internet answer these questions. They produce a new image of nature. This image truly takes on form when considered within the context of a parallel development that began with the Information Age and continues to revolutionize our perspective on society. The concept we have of the structure of society has changed radically in the past ten years. The image of social strata and milieus, even of nationalities, used to be dictated by ideological default and assignation. The image associated with the working or middle classes used to align with the political theories that both engendered. "The Frenchman" or "the German" more or less represented national clichés. We primarily saw what we knew. We would then acknowledge representatives of these classes and nations and judge them accordingly. The ideologically fixed gaze first began softening with the introduction of new modes of direct communication and observation that established a new realism in perception. Telephone and television combined first with the Internet, then with social media, offering a visually supported opportunity for exchange and hypothetically providing any user with direct access to the unique lives of other users. The homogeneous, cut-and-dried preconceptions of class and nationality have increasingly fallen away as a result. Social media provides us with a picture of society as it truly is: full of contradictions, intersections, and dissonance. Today, those who want to hold tight to their ideological perspectives need to work a lot harder at it, and all too often, their efforts collapse when faced with the temptation to gain concrete impressions of the world. These new insights generate a dynamic that is also changing society. Because seeing creates knowledge, and knowledge leads to action. The social web is prepared to deconstruct social theories from the inside out, by starting an engine of social and systemic change free from ideological theorizing, as recent events in Egypt, Tunisia, Turkey and Ukraine have shown. Mobilizing the masses and coordinating social processes—these were the central functions of ideology. Social media has now taken on these duties. The Net has brought about a paradigm shift—out with theory and ideology, in with practice and reality.

The Human Internet has changed society, and the Animal Internet will change nature. We speak relentlessly about how modern technology has influenced human communication and interpersonal relationships. Since the birth of the Internet of Things (IoT)—that is to say, since furnishing inanimate objects with intelligent sensors, making these things trackable and "sentient"—the conversation has expanded to address the consequences of this technological revolution on humans' relationship with their inanimate surroundings and on our society as a whole. It is no longer just humans who can use the Internet, sending and calling up data, but also devices, switches, and sensors that can be connected to the Web and interact without the need for human involvement. Packages that can be tracked by means of integrated electronics and printers that automatically order replacement cartridges when they are running low are innocuous examples of the IoT. The same goes for fitness bracelets, electronic pedometers, and the handy new features of the connected home. But rooms with sensors that register human presence and then identify these humans and match them with data pulled from the Net hold greater potential for risk—even if there are certainly many helpful possible uses for so-called "smart space" technologies.

What about when intelligent technology allows for not only things to start thinking and speaking, but animals, as well? What happens when wild animals start pinging us, and we are able to identify them as unique individuals with their own backstory? The discussion on the ways in which digital technology can reshape our relationship with other living creatures and with nature is still new; in fact, it hasn't even really begun. We can, however, already foresee the revolutionary effects of the Internet on our awareness and knowledge of nature. The technology exists to allow animals to communicate autonomously, thereby affording us a realistic impression of nature that deviates from the necessarily ideological picture that two hundred years of natural history writing and ecological theory have painted. This new conception does not derive from theory—from Darwinism, behavioral theory, the notion of an ecological niche, and so on and so forth—that claims confirmation through a single concrete observation; instead, it emerges from a glut of data and information. The focus is now on the individual animal, rather than on confirming the individual class. In the animal kingdom, our attention is no longer paid to family, genus, or species, having shifted instead to the individual with its specific history.

The animals of the Animal Internet are not user-generated content; they are not memes, those packets of digital information that spread at the speed of light and have given rise to a new visual "culture." Instead, they themselves generate and transmit data. Animals and even plants—like essentially inaccessible trees in the rainforest, the growth of which can be measured by so-called dendrometers—are equipped with sensors that transmit information about them, and not just about their movement, but also various environmental data (temperature, air pressure, etc.) and physiological readings from the animals' bodies. Many animals are already tagged with powerful GPS devices on or even in their bodies: snow leopards, humpback whales, albatrosses, red-eyed tree frogs, fruit bats, ocelots, saiga antelopes, hammerhead sharks, orchid bees, mountain gorillas, storks, and brown bears. These transmitters make it possible to follow the animals, no matter where they are: in the heart of the rainforest, ranging the desert, or even far below the sea, and thanks to Internet technology, we can access this information from anywhere on Earth. More wild animals are outfitted with sensors every day. As a result, a huge store of data is gradually coming together to form a nuanced and differentiated image of nature that will ultimately serve as a complex portrayal of animal life.

The Animal Internet is unquestionably a technological revolution. At the heart of the Animal Internet are miniscule transmitters powerful enough, even, to send information into space. In the first half of 2016, the International Space Station (ISS) will install a special antenna to receive these signals. Biologists anticipate an exponential increase in information. The signals, weakened from the distance traveled, will be processed at the space station and sent to a database, where the information will be translated into visuals. The ISS antenna is intended to accommodate around fifteen thousand receivable transmitters. Future plans include mounting antennas to low-flying satellites, to achieve yet wider coverage and the ability to follow animals in real time. This will then someday allow us to tag classes of smaller creatures, like insects (which represent the largest number of species in the animal kingdom), and follow them individually from space. After all, even the migratory patterns of butterflies remain largely unexplored. This is naturally not the solution to all the problems. A key challenge is powering the transmitters, which must operate without an electrical connection. They will all run out of power at some point, since batteries cannot simply recharge themselves. Through the development of intelligent, energy-efficient technologies, the Animal Internet is giving rise to a more tight-knit structure that unites macro- and micro-perspectives.

Four components define the system behind the Animal Internet: the first step consists of tracking the animals. This is followed by data transmission to mobile phone networks and Internet hot spots or out into space. Then the third component, a database located at Movebank.org, receives and processes the data. Finally, the visually formatted data can then be presented to scientists, laypersons, and hobbyists in mobile apps such as Animal Tracker. The analysis of movement and behavioral data offers insight into a multitude of problems within the fields of theoretical and applied biology, the solutions to which were long out of reach, because of a lacking or limited base of knowledge.

GPS units are particularly well suited for tagging wild animals because they can be pinpointed from long distances, which the practice of classic telemetry, requiring researchers to drive after the animals with a transceiver, did not allow. The transmitters have become so refined and so small that it is now possible to track many animals for months or even years. The transmitters should not weigh above five percent of the animal's body weight, which presents a great challenge to researchers and their technicians, because it means a transmitter for a .70-ounce chickadee cannot weigh over .035 ounces. Transmitters have already been built, however, that weigh just .0007 ounces. This enables even insects to be tagged. How far does a bumblebee fly to reach its food? What is the radius of its movement? This was previously unknown, until a team at the Max Planck Institute on Lake Constance in southern Germany used transmitters to get to the bottom of these questions and discovered that bees will fly several miles to reach their food sources.

A transmitter implanted in the body of a wild animal is naturally a great disruption to the animal's life and may impede its mobility. As light as the unit may be, the danger remains that it will limit the animal's ability to move, along with its chances of survival. Attaching it to the animal's body is also no mean feat. It is almost impossible to imagine sewing a transmitter onto the three-inch-long body of a red-eyed tree frog, which has the slippery skin typical of amphibians, without injuring the creature. This procedure must be tested repeatedly under controlled laboratory conditions before it can be implemented in the wild. The prototypes of each new generation of transmitters are designed such that, should they interfere with key life activities such as reproduction, the animals can easily free themselves of the devices.

The form the data transmission takes, whether permanent or piecemeal, depends on local communication infrastructure. In areas lacking infrastructure or when tracking animals over the course of a long migration, the data transfers occur via satellite; the data are fractionated, meaning they are collected and sent intermittently as bundles. In order to do this, the information must be stored temporarily on chips. A big technical hurdle at this point is how to supply the chips with power. Different types of batteries come into play, ranging from high-performance batteries to solar cells to kinetic systems. A key concern is efficiency, because the batteries are tricky to replace. It really comes down to the intelligence of the chip design. Chips can be programmed to be active only at specific times. Or they can be controlled remotely to turn on or off, bundle data, or upload data at a scheduled time. It is also possible to process the collected data on the chip itself, and to send only the results. Text message responses can even be programmed and stored on the chip, then sent automatically upon receiving certain signals.

This technique is frequently used in western Australia to warn swimmers and surfers of sharks approaching the coast that might pose a threat to humans. This applies primarily to great white and tiger sharks, over three hundred of which have already been tagged. In order to track them, the creatures are first caught, tranquilized, and pulled on board. Remaining at sea, marine researchers then perform a quick operation to implant a small transmitter into the sharks' abdominal cavities. Since radio waves travel poorly through water, the devices emit sound waves that are picked up by underwater microphones. Whenever a shark swims within range of a microphone, it logs in with an individual identification badge. The signal is then forwarded to a network of monitoring stations. These data provide important information about the animals' migration patterns. The moment a shark crosses one of these so-called digital "geofences," its arrival is announced via text message or Twitter. Signals also travel via satellite to monitors installed on beaches.

Another, somewhat spooky example of a tagged shark demonstrates the chasms GPS technology will reveal to us in time. Off the coast of Australia, a ten-foot-long, tagged female great white shark that went by the name "Shark Alpha," disappeared from the radar. According to the tracking device, at four o'clock in the morning, the shark was suddenly torn five hundred yards into the depths, with astonishing power and speed. Within seconds, the chip also recorded a spike in ambient temperature, from 8 degrees Celsius to 25. That is the temperature of an animal's insides; the shark must have been eaten by an aquatic predator. The chip could be followed for the next eight days, at which point it vanished from the control monitor. It was most likely voided. Four months later, it was found on shore, bleached by gastric acid. Researchers suspect Shark Alpha fell prey to a much larger creature. It will have to have been at least five meters long and weighed two tons or more. But what was it? An orca? Orcas usually hunt close to the surface. The deepest killer whale dive on record is 260 yards. Another great white? This species has a body temperature of 18 degrees—not 25. Could it actually have been a monstrous octopod or a megalodon, a gargantuan prehistoric predator that some say may have survived, hidden in the darkest depths of the ocean?

As this example illustrates, the data collected need not be limited to sending updates on an animal's position. The chips can also relay information about surrounding conditions, from climate data to air and water pressure. In some cases, specialized sensors can provide readings on an animal's general physiological condition. The primary foci may include heart rate, body temperature, and blood sugar levels, but more complex bodily functions may also be captured by means of EKG, EMG, or EEG tests. Researchers can thus determine from a distance whether an animal is sick. Not only does this technology improve the care wild animals receive, it also better enables scientists to determine the prognosis of spreading diseases or even epidemics. Finally, these data may be combined with audiovisual information to convey the most precise possible account of each animal's current situation; these accounts then serve as the building blocks of a realistic overview. With this in mind, Martin Wikelski, director of the Max Planck Institute for Ornithology and one of the Animal Internet's leading zoologists, is planning to equip birds' beaks with tiny cameras that are triggered by characteristic head movements during feeding. This would allow the animals' daily menu to be recorded in high definition.

With such precise and readily available data, nature researchers are no longer dependent on supposition, deduction, or their own imagination; the impact such an informative, objective image of nature will have on conservation campaigns is evident. After all, these data provide crucial insight into migration routes and population sizes, as well as into habitat issues and possible areas of conflict with humans. They help answer questions that have long been unanswerable.

MEANDERING ROUTES THROUGH THE DARK NORTH PACIFIC

A glaring number of questions remain about the unknown lives of animals, but with the help of digital technology, researchers are getting to the bottom of them and using their findings to help improve animal habitats and living conditions. Tracking animals that live underwater still presents some of the greatest challenges. For those who used to rely on assumptions about the tortuous routes elephant seals take through the dark North Pacific, however, it has been possible to hitch a ride on "Lady Penelope's" hind flipper and share in her experiences. This impressive representative of the world's largest seal species greets visitors to her Facebook page—which users have followed from the *Tagging of Pelagic Predators* (TOPP) website—with a warm, "Hi, my name is Penelope, my home beach is Año Nuevo, and I have swum a distance of 8,910.04 miles since I was tagged."

Elephant seals are especially well suited to tagging, because they always return to the same beach. The chips can then be changed or read. Since elephant seals travel far distances in the ocean, they're interesting objects for study. Researchers can collect a lot of data from their extended travels. Thanks to continuous digital monitoring, Penelope's life story can be recounted from start to finish. In addition to charting her movement, her chip records the depth and length of her dives, as well as variance in light. Her august history reads as follows: Born in the first half of January 1998 in Año Nuevo State Park on the coast of California, she weighed but a dainty forty-six pounds. Today, the noble creature registers fifteen hundred pounds on the scale. All her life, and like a true aristocrat, Penelope has disdained the sensibilities of the petite bourgeoisie, rejecting conventions of monogamy, and unreservedly seeking the company of multiple partners. She lives in a so-called polygynous society, in which she shares her mate with other females. Fidelity has never been her strong suit, though, and she has been known to socialize with other beta males from time to time. She has six children, the first of which she had at age five. This carefree lifestyle has paid off—at age twenty-six, Penelope can be considered one of the great survivors of her species. As it happens, fifty percent of these animals die before reaching maturity. Meanwhile, Penelope has taken an even greater stage: she was recently integrated into the Oceans Street View in Google Maps. Her travels through the ocean yield pictures of regions that are then added to the virtual underwater atlas.

Since 2000, the TOPP project has connected marine researchers from all over the world. In this time, the zoologists have managed to tag twenty-two different species—elephant seals, great white sharks, sea turtles, squid, tuna, albatrosses—and over two thousand individual animals with satellite transmitters. From its inception, the program has aimed to make these scientific data available to the general public. One of its stated goals is to lift these animals from obscurity and share their life story. This push to "let nature tell its story" is a central feature in the new era of nature conservation. It draws on the power of story. On the one hand, scientists are using technological tools like websites with integrated blogs for individual animals, or smartphone apps like Shark Net that allow users to select specific sharks they want to follow. On the other hand, the researchers also do not shy away from crowd-pleasing events like the annual Elephant Seal Homecoming Day, at which the tagged animals returning to their home beach are greeted with beers and barbecue; or the Great Turtle Race, in which various sea turtles compete in a virtual race to the Galapagos that is broadcast online for fans to follow. The primary goal of these formats is to get users and viewers to adopt the animals' perspective in a playful manner that creates a sense of connection. Nature can be entertaining again.

Nevertheless, the gravity of the situation keeps returning to the forefront. Take, for example, the Magellan penguins that move along the Argentine coast. These animals often encounter oil slicks when swimming through shipping lanes. When the oil comes into contact with their plumage, the penguins can no longer maintain their body temperature, and they die of hypothermia. Those that survive are plagued by health issues and can no longer reproduce. The oil pollution along the Argentine coast kills as many as forty thousand penguins every year. P. Dee Boersma, one of the world's leading penguin researchers, started tagging penguins with GPS devices as early as the mid-nineties. This allowed her to determine the penguins' usual migration routes. She then used these data in negotiations with the Argentine shipping authority, which agreed to move its lanes farther out to sea. This series of events drastically improved the survival rates of this penguin species, if not, in fact, protecting it from extinction.

WHY WE SHOULD CARE IF A FROG WANDERS AROUND IN CHINA

THE NEW GENERATION OF WORKING ANIMALS

Truffle pigs and Seeing-Eye dogs—not many more examples come to mind when prompted to list the ways in which humans today use animals' unique skills. The animal sensorium is, however, much more highly attuned to its surroundings than the human equivalent. Were all the earth's animals brought together in one place, they would create a sensory supersystem far superior to any human capacity or creation, including the newest technological inventions. But how can we tap into this system? How can it be used to address the unresolved problems—geological, climatic, or medical—facing humans? The emergence of a new generation of "working animals" is one of the most interesting aspects of the Animal Internet.

Something surprising happens the moment we start to collect and systematically analyze new knowledge of animal behavior and mobility: we recognize that the process of generating knowledge is not one-dimensional, that it's not just about devoting time to animals and helping them survive, thanks to our newfound wisdom. Our increased knowledge of animals has other implications: it translates into insight into our own limitations. After all, the new facts gleaned from the animal kingdom contain unimagined, vital information that can be used to solve key problems facing humanity. This approach, contained within a discipline that could be called "translational zoology," confronts us with the fact that not only do we have a lot to learn *about* animals, but that we also have a lot to learn *from* them—and that zoological knowledge can be translated for human use.

Over the course of evolution, the animal world has amassed a trove of experience inaccessible to us, but that we can now access with the help of technology. Not only does the new image of nature broaden our perspective on the natural world, it returns us to a state of humility we thought we had put behind us, given our technological world domination. Over the course of Earth's history, animals have conquered every last corner of the planet. However extreme the given environment, animals have developed the strategies needed to survive, establishing biological niches in the harshest of conditions. Animals have colonized barren deserts, lightless caves, tropical rainforests, and Arctic regions. Their alertness is unrivaled, both physically and temporally. And not only because they can be both diurnal and nocturnal: the animals living on Earth today have seen the world from all angles by both day and night, and they have lived through ice ages, natural disasters, and much more. This has made them into specialists. A collective sensory repertoire has developed in the animal world over the course of evolution that allows animals to recognize and contextualize signals and stimuli, such as instances in which they will flee long before humans have perceived any approaching threat.

An individual animal's behavior in a dangerous situation is not a deliberate, one-time decision; rather, it is based on millions of years' worth of accumulated behaviors. The aggregate wisdom of the animal world—and "wisdom" is defined here as the store of experiences in which precise interpretations of the outside world's sensory stimuli were introduced and refined—has been encoded and conserved in the genome and phenome, as one might say, of the animals living today. Humans of all cultures and eras have relied on animals to warn them of impending danger or to help locate food sources. In ancient civilizations, animals were always indicators of specific situations: they signaled danger, water, prey, changes in the weather, or other catastrophes. Many peoples were convinced that animals possessed a "sixth sense" for natural phenomena, and that particular behaviors or, indeed, their very appearance was a sign, an omen for a significant event coming to upset the course of everyday life. The Etruscans and ancient Romans viewed any large influx of certain animals—like snakes, wolves, or owls—as an ominous sign of political significance. Omens applied to the polity as a whole and were handled accordingly: they were formally recognized by the senate, and the Romans performed collective rites of atonement, in efforts to appease the gods. There are certain areas in which people today still make use of the animal sensorium. Dogs are used to sniff out drugs. A species of songbird was even adopted for a very specific purpose: coal miners would take canaries underground with them, and if one of the birds stopped singing, it was a sign of poisonous gases or low oxygen levels.

Countless miners' lives were saved as a result.

But even those days are gone now. Today the animal sensorium lies fallow. The functional relationship has been razed. This is the terminus of two hundred years of evolution, a chapter that features not only the collapse of a pragmatic relationship, but also the dissolution of the "existential dualism" of human and animal—the mutual entanglement—that arose from this relationship. This history of a productive interdependence and its abrupt end illustrates what the Animal Internet can build upon and from where it originates.

Until the end of the Second World War, animals lived in direct proximity to humans. They practically lived on top of one another. As the English essayist John Berger described in a seminal text on human-animal relationships, animals formed humans' inner circle. Animals shared in daily human life, and not only in the country, but from the time of the French Revolution onward, in cities, as well. They sometimes slept under their owner's roof, in the bed, even under the blankets. They were productive helpers in human society, providing modes of transportation and pulling farm equipment, guarding houses and villages, and supplying food and feedstock. Animals were an essential part of human life, which is why people felt an emotional, but far from sentimentally idealized, connection to them. Life without animals was simply unimaginable. Cathedrals could not have been built, nor wars waged, without animals. In the First World War, the first truly technological war, a close symbiosis still existed between humans and animals, as depicted in Stephen Spielberg's powerful 2011 film *War Horse*. Dogs were used to transport the injured and deliver messages, lay telephone wire and lead blinded soldiers back from the front. Some animals were even decorated for their service, such as carrier pigeons that were employed when radio contact was not an option.

The last two hundred years are thus central to understanding the human-animal relationship. As it happens, in response to the pressures of burgeoning human populations and economies, the number of animals in Europe during this period grew substantially. Initially the engine driving industrial development, animals would ultimately fall victim to the same. Existential dualism also intensified during this phase. At the very moment in time animals were helping advance industrialization, they were also losing ground in the everyday world of humans. Mechanization made animals redundant. There was a reason the term "iron horse" was popular in Victorian England and the United States.

The opposing and parallel developments of animals' disappearance and humans' desire to cling to them are interesting. The doggedness with which humans bound themselves to animals, even if it meant behaving anachronistically or risking danger, is astonishing. This doggedness was certainly also in response to economic concerns. At the start of the twentieth century, there were many who could not afford a car or even a train ticket. It would, however, also appear as though human companionship with animals provided some protection against the inhospitable structures of economic modernity, which was penetrating steadily into the folds of daily reality.

Humans launched a countermovement to combat the disappearance of animals, as the numbers show. The number of animals in human society increased dramatically between the end of the eighteenth century and the early twentieth, before dropping rapidly from the 1940s onward: while animal ownership may have been limited primarily to rural areas through the mid-eighteenth century, their presence in urban and suburban areas rose steadily from the time of the French Revolution onward. French figures illustrate this growth clearly: while seven million cattle existed in France in 1789, 14 million roamed the fields in 1914. One million dogs lived in France before the revolution, and a century later, the count was up to three million. And despite the invention of the steamship, railroad, and the automobile, which changed the face of human mobility and made life more comfortable, the number of horses used in everyday life grew steadily through the 1930s from two million to three. The number of animals in France doubled, or even tripled, during this time period, while the human population grew by just 45 percent.

Between the French Revolution and the Second World War, there are thus two opposing movements at work, the intersection of which led to moments of extraordinary juxtapositions that manifested themselves, for example, in the cavalry attacks against rapid-fire machine guns during the First World War, or in the dog-drawn milk-carts used in Paris well into the days of the automobile. While on the one hand, humans clung to animal companionship as long as they possibly could, even when better alternatives became available—alternatives that allowed work to be accomplished more quickly and involving less risk—mechanical constructions gradually came to replace animal helpers. This drive to hold fast to animals as companions in work and everyday life is evidence of the linkage between humans and animals. By the end of the Second World War, however, this process, which was thousands of years old and had significantly shaped humans' worldview, ended abruptly.

Up until that point, the proximity of animals had expanded humans' sensory radius. Human and animal senses were synchronized. Through contact and close observation of animals, humans learned to explore their physical surroundings, recognize opportunities for themselves and their communities, and avoid approaching risk well ahead of its arrival. When animals fled, danger was near. When swallows flew lower to the ground, a storm was nearing. When horses shied or frogs jumped out of ditches, an earthquake or other natural disaster was imminent. Goats streaming down Mount Etna were a sign that the volcano was about to erupt. According to ancient belief systems, animals must be in the know, because they are protected by the gods: in Ovid's fable of Philemon and Baucis, *"unicus anser erat, minimae custodia villae"*—"there was a single goose, the guardian of their little cottage"— that was intended to be killed in honor of the gods Zeus and Hermes, who had entered the house in disguise. The code of hospitality demands it, Philemon and Baucis believe, but the gods prevent them from killing the goose. They protect the animal. In honor of the couple that has shown such hospitality, they transform the cottage into a golden temple, grant them a joint death, and change them into an oak and a linden tree. It is impossible to read the end of the tale of Philemon and Baucis without feeling profoundly moved, even after the hundredth time. A magnificent and deeply affecting order is thus preserved, an order that is closely tied to nature, at the heart of which is a silly goose, and in which eternity passed in the form of a tree is depicted as the ultimate reward.

Humans now trust almost nothing except technology. When today's humans want to observe and assess their surroundings, they no longer rely on animal instincts (not even their own), but on gadgets. They record data and install cameras. Although a long tradition exists of the pragmatic interpretation of observable animal behavior, animal sensory capabilities have not yet been implemented in the systematic mastery of central problems facing humanity. One reason for this is the difficulty in reliably gathering and analyzing information.

ANIMALS AS EARTHQUAKE DETECTORS

In the past, however, reports of animals' almost wondrous-seeming sensory capacity received a lot of attention. One famous example is the Chinese city of Haicheng, which experienced an earthquake in 1975 measuring 7.3 on the Richter scale. In response to animal movement, the city was evacuated in time, and many residents survived the quake without damage. Chinese officials subsequently distributed informational brochures to raise people's awareness of alarming animal behavior. In Sri Lanka and India, elephants have been known to flee in panic before tsunamis. In similar situations, flamingoes will abandon their breeding grounds, and water snakes and amphibians also seem to be good indicators of plate tectonics. Following the mass movement of toads observed in advance of earthquakes in China and the 2009 L'Aquila earthquake in Italy, scientists have been paying closer attention to this connection. They have formulated the hypothesis that, preceding an earthquake, particles in the earth's crust loosen and mix with the groundwater. Animals that live in or near groundwater, such as frogs and toads, are very sensitive to such changes in their immediate surroundings. The L'Aquila earthquake allowed for a particularly close analysis of this connection; a biologist had been studying a local colony shortly before the catastrophe, and she could therefore describe precisely the concrete connections between the seismic activity and the creatures' movements. Birds also have an extremely fine-tuned sensorium for potentially dangerous changes in their surroundings. Pilots have observed the way in which bar-headed geese will sense impending avalanches and increase their usual flying altitude by about six thousand feet in response.

Nature never stands still. The commonality underlying these examples is the fact that movement patterns serve as a source of information. Animals often communicate by means of their movements, in that they respond to changes in their surroundings by changing their location. Were it possible to collect and analyze these mobility models systematically and comprehensively, using tracking technology, new methods would emerge to mitigate the deleterious effects of many key dangers. Animal movements could warn us of sea- and earthquakes, thereby significantly increasing the reaction time in vulnerable areas. In the instance of a tsunami, twenty minutes longer forewarning would result in a drastic decrease in casualties and a reduction in resultant costs of up to 30 percent. During this window, power plants could be deactivated, people evacuated, and building complexes cleared. There are also less dramatic-sounding contexts within which the "disaster forecasting" made possible by the Animal Internet would be beneficial. Climate change is sparking global animal migrations with dangerous potential. Animals are encroaching on areas they have never before inhabited. In doing so, they may ravage local flora, negatively impact the food chain, or spread disease to humans. The ability to forecast these new mass migrations would greatly aid in assessing the associated risk to the food, health, and environmental systems in affected regions.

Pilot projects have already been launched to test this new method of gathering information. In 2011, Martin Wikelski, of the Max Planck Institute for Ornithology, tagged goats on Mount Etna to study their movement patterns before explosions or volcanic eruptions. It was the first time that GPS transmitters were used to study animal behavior *before* catastrophic events occurred, to see if animal behavior changed significantly in the time preceding seismic activity. The animals were thereby transformed into biosensors. The initial results have been encouraging: up to five or six hours before larger volcanic events, the goats noticeably change their movement patterns, speeding up erratically or relocating to an atypical destination, for instance. Another such project could use frigatebirds to detect hurricanes in the Caribbean. Upon observing the birds' instinctive patterns of evasion before an approaching storm, humans could draw conclusions about the location of safe zones.

ANIMAL MIGRATION AS DISEASE CONTROL

The Animal Internet can also be used to better study the spread of highly contagious and deadly diseases. The path viruses take when carried by migratory animals is now more easily predicted, allowing us to implement precautionary measures more quickly. The Ebola outbreak in Guinea serves as a good example. Fruit bats are recognized hosts of the deadly Ebola virus. Transmission is believed to occur primarily through the consumption of tainted monkey meat, the monkeys themselves having eaten fruit contaminated through contact with fruit bat excrement. Furthermore, in West Africa, fruit bat meat is also considered a delicacy—especially because it is believed to have restorative qualities after excessive alcohol consumption. There are a lot of these animals, and they are easy to catch, which makes them an affordable food source. The furry bats are cooked whole and served in a peppery soup. Fruit bats fly long distances. The straw-colored fruit bat (*Eidolon helvum*), for instance, might fly over nine hundred miles. These bats are believed to be a keystone species in the rainforest, because the animals' excrement contains the seeds of untold plants. Upward of 96% of trees are believed to have sprouted from the dung of fruit bats. Most frugivores live in the treetops and defecate from the same spot. Not the case with fruit bats. Their delight in flight is as pronounced as their bowel movements are infrequent. They regularly undertake sixty-mile flights to reach their next feeding spot. Sometimes they even turn around and fly straight back, to give their digestion a nudge. These creatures pose a threat to humans as disease carriers, particularly given their mobility, coupled with the irregularity of their defecation. Because of their wide migration radius, fruit bats can spread dangerous viruses over great distances. If it were possible to track these animals' migratory patterns, it would be easier to predict the spread of a future Ebola epidemic and contain it more effectively than is currently the case. At present, the sole method for containing Ebola is military isolation of impacted areas. The fruit bats don't let that stop them. With the help of tagging, even our native mallards could be useful in the fight against epidemics, by providing early detection of bird flu. The devices on their bodies record heart rate and body temperature in real time and transmit these data to the Web. It is thus possible to identify and locate individual diseased animals, stopping the outbreak of a dangerous epidemic before it starts.

CRITICAL QUESTIONS REGARDING A TRANSPARENT NATURE

These examples illustrate limited scientific pilot projects intended to show the range of possibilities this approach could provide. In order to obtain statistically reliable reports on the spread of diseases or the approach of natural disasters, a staggering number of animals would have to be captured, tagged, and followed. Besides, without satellite monitoring from space via the International Space Station, nothing can be done. Within the scope of the revolution started by the Animal Internet, the use of animal experience to solve human problems plays a secondary role. It does not figure into the central benefit promised by the new, communicative nature. This benefit can be found in the changed relationship between humans and animals. Despite the many opportunities for innovation, critical questions regarding an increasingly transparent nature arise. Although sensory technology is already in use and research is progressing rapidly, the public debate—when it actually occurs—over the resulting upheaval cannot keep up with the pace of events, and always seems to take place after the fact. It's imperative, however, to consider the consequences of this new, systematic utilization of animals. For instance, were these projects to be developed into wide-scale early warning systems that could then be sold to private entities, such as insurance companies, to aid in risk calculations, the impact on nature would be considerable. Is it ethically justifiable to catch and tag huge numbers of animals, in order to connect them to our network and exploit their sensory capacities? Can we defend the stress unleashed upon animals in this way because of the potential benefit to humans? What are the limits to animal digitalization of this magnitude?

It would of course be desirable to predict as many natural disasters as possible, as accurately as possible, in order to save human lives and infrastructures. But there is a danger inherent to the Animal Internet, namely that through its use, humans will again appropriate nature for their purposes—subjecting it to their will and pursuit of gain, subjugating animals once and for all. The possibilities are far from exhausted. The next question may very well be if humans will decide whether any given species is justified in its existence and therefore worth protecting based solely on the criterion of usefulness to humans. The only species then worth protecting would be either those of highly symbolic value through their genetic closeness to humans, such as primates, or those that are especially relevant within the context of sensory-driven utility. From an economic standpoint, conservation budgets would focus on these species. In combination with the possibilities of synthetic biology, this danger could be situated in any number of constellations, in which extant species are genetically modified to better perform their duty as bio-indicators. A bestial brave new world is on the horizon. Keeping this vision in mind, one of our top priorities must be providing an ethical basis for the Animal Internet. In Edward O. Wilson's *The Future of Life*, a book which proposes ways out of the conservation dilemmas the world currently faces, he underscores this point: "The new strategy to save the world's fauna and flora begins, as in all human affairs, with ethics." Ethics are not a question of definition, though; rather, they are a question of personal responsibility. Those responsible for the Animal Internet are, at least initially, its inventors. Martin Wikelski is one of them.

WHY ALEXANDER VON HUMBOLDT HASN'T LOGGED OFF YET …

THE PEOPLE BEHIND THE ANIMAL INTERNET

Alexander von Humboldt, born in Berlin in 1769, was the son of a Prussian officer. He and his brother, Wilhelm, two years his senior, grew up together in Tegel Palace, in north Berlin. The younger von Humboldt brother, who wanted to become an explorer, studied physics and chemistry, economics and foreign languages, geology, anatomy, and astronomy. An inheritance allowed him to realize his lifelong dream. In 1799, von Humboldt traveled through South America, discovered a connection between the Orinoco and Amazon Rivers, crossed the Andes, and scaled the Pichincha volcano (15,728 feet) with the French explorer Aimé Bonpland, making them the first Europeans to do so. They had to halt their ascent of another volcano, the Chimborazo (20,564 feet) at 18,700 feet: Humboldt was the first to describe the symptoms of elevation sickness. He spent another year in Mexico. After being received by Thomas Jefferson in the new capital city of Washington, Alexander von Humboldt returned to Europe in 1804.

During his time in South America, von Humboldt

identified sixty thousand plants and discovered sixty-three hundred new specimens. He collected important geological and geographical findings, wrote political treatises against slavery that were widely disregarded in Europe, and fired the imagination of European speculators by writing about silver mines in Mexico in one of his studies. Between 1807 and 1833, the findings of his South American expedition were published in thirty-four volumes in French. He returned to Berlin in 1827 and delivered lectures at the university his brother had founded. Von Humboldt, who lived to almost ninety and died in 1857, had also traveled through the Ural Mountains and Siberia. *Cosmos: A Sketch of the Physical Description of the Universe* (1845-1862) is widely considered to be his magnum opus, where his multifaceted interests coalesce. His research was always focused on the big picture, because he suspected a universal connection between earthly phenomena. He fostered interdisciplinary exchange and was an intellectual pioneer; indeed, he was a pioneer of globalized science. In a sense, Alexander von Humboldt can also be thought of as a pioneer of the Internet—and the Animal Internet. After all, he was thoroughly convinced, not only of the idea of interconnection, but of the necessity for the democratization of knowledge.

But in his wildest dreams, of which there must have been many, could Alexander von Humboldt have imagined that it would ever be possible to identify individual whale sharks using what we know about the stars above? This can be done today. Whale sharks' spotted skin is reminiscent of the starry night skies. Every shark has its own distinctive pattern. The arrangement of white marks on the creature's body functions like a fingerprint; it verifies each shark's unmistakable identity. The sharks are tagged virtually using these patterns. They are identified using an algorithm originally developed for the Hubble Space Telescope to locate stars and galaxies through pattern recognition. This software allows individual whale sharks to be identified by snapshots that divers upload to Facebook. Using this data, whale sharks can then be tracked as they move through the oceans, helping researchers investigate many important aspects of this highly endangered species' lives.

Some might ask, what kind of animal researchers spend their lives comparing the skin of a whale shark to the Milky Way, or catching frogs and sewing trackers to their bodies, instead of directly confronting nature and gathering observations using established practices? Isn't that part of our romantic image of the natural scientist, sitting for hours in a camouflage tent, waiting with the patience of Job for the moment the animal finally ventures out of the thicket? At first glance, those who mess around with technology to follow red and blue signals on a monitor do not have much in common with the likes of great naturalists such as Konrad Lorenz, Karl von Frisch, or Bernhard Grzimek. Instead, they seem more like engineers—enemies of nature, even— and sober technocrats, than like those who have vowed unconditional friendship to animals. In order to provide just that, however, today's naturalists must avail themselves of wholly different methods than even twenty years ago. Animals are put under increasing pressure because of negative human impact on their habitats. We can no longer embrace the "live and let live" attitude, the dream of leaving nature to its own devices, and of its closing itself off to humans, deep in the Amazon or on the mountaintops of Nepal. To continue to do so would be a tragic mistake. Wild animals must be protected by humans. The objective must become to safeguard their coexistence in surroundings increasingly shaped by humans. And this can be accomplished only through technologies that do not leave an animal's whereabouts and condition to chance, but that make it possible to monitor its life constantly. Today's zoologists thus use technology out of a pure, unqualified love for animals. It serves as a medium for them to gain true ground with the wild. The new image of nature is preceded by a new breed of naturalist; these scientists are in possession of a wholly new set of competencies and interests. They don't see nature and technology as diametrically opposed; instead, they believe that we must employ a new discipline to establish the new image of nature we need. At the same time, modern-day naturalists are substantiating Alexander von Humboldt's conjecture that an essential connection exists between all living things—an idea that von Humboldt was pursuing along speculative, interdisciplinary, and literary lines, but that he ultimately lacked the tools and other resources to prove. Von Humboldt could never have envisioned the technological aids in use today, so instrumental in verifying his vision of a universal connectivity.

How, though, does the question of the interconnectedness of all living things, which dominated Humboldt's thinking, translate into the here and now? What questions would Humboldt pose today?

He would probably focus on three new connections: first, a deeper understanding of the planet's biodiversity, from its constituent parts to the connections between those parts. Second, the connection between us and the trove of animal experiences amassed over the course of evolution. Third, the digital storage of every last species in a virtual cache, a global "bio-cache," from which an entirely new human-animal relationship can emerge.

Martin Wikelski could be seen as a Humboldt of the Digital Age, the prototype of the modern animal researcher. Wikelski is as confident within the aerospace and telecommunications industries as he is studying the social behaviors of fruit bats and frigatebirds. In order to understand the long road zoology has traveled in the last fifty years, we must keep in mind that in his role as director of the Max Planck Institute for Ornithology in southern Germany, Wikelski is Konrad Lorenz's successor. Despite their many differences, the two researchers are bound by an early childhood interest in nature. Wikelski describes the beginning of the personal journey that would lead him to the Animal Internet as follows: "I grew up in a small farming town in Bavaria, which meant I was in constant contact with animals and the natural world. As a boy, I was allowed to take my grandfather's cows to pasture and then bring them back in. As long as I can remember, I knew I would choose a profession working in and with nature. The decision to study biology came in the fifth or sixth grade, because I had a very good biology teacher who, among other things, helped me understand how a group of cattle egrets from the tropics had ended up in Bavaria in 1977." To this day, Wikelski is motivated by the questions that accompanied him through youth: "I would still like to know where the swallows go, like those that I first banded as a fifteen-year-old in my grandfather's barn. The next year, they returned to our town from Africa, and in pretty high numbers, at that. Where they fly, what happens to them, what difficulties they face—this all remains one of the greatest unsolved mysteries."

There are countless reasons why humans concern themselves with animals. Noting the presence or absence of animals has been a constant. Animals wander, move, and migrate. They are always in motion, in search of food, breeding grounds, or hospitable climes in which to winter. Humans whose existence depended on animals, like hunters or shepherds, needed to understand animal movement patterns. This formed the foundation for our fascination with animal migration. Martin Wikelski is part of the search for the answer to this riddle, which has kept people wondering since the days of Aristotle and which is still difficult to solve, because observations of the animal world remain limited to chance and isolated connections. It's impossible to use conventional methods to track animals—particularly smaller animals, like insects or birds— over long distances without gaping interruptions.

Initial attempts to track animals were made employing banding, the practice of applying small tags to individual animals so they could be identified when spotted and caught. Banded animals' movement over long distances could be recorded, but this tracking was subject to chance. The introduction of targeted technology closed many of the holes inherent to this older method. Telemetry, or wireless data transfer, was an important step forward in this regard. Wikelski strongly believes that zoologists must keep trying new technologies in order to make substantial advancements in their findings: "To me, technology has always been the key to understanding new things. As a child, I bought a tape recorder, which was unbelievably exciting at the time, and later a microscope to observe the microorganisms living in the waters of Lake Worthersee in Upper Bavaria, which gave me insight into an entirely new ecosystem. Not long after, I had the opportunity to use an electron microscope at the Bavarian State Department of Water Resources, which allowed me to delve that much more deeply into the fine structures of brown algae and other organisms, an experience that more or less bound me to technology for the rest of my life."

This was Wikelski's motivation in founding the ICARUS Initiative in 2002. ICARUS stands for "International Cooperation for Animal Research Using Space." This project, based at the Max Planck Institute, is working with its partners worldwide, including all relevant space agencies, toward establishing a global monitoring system for small migratory species. The interplay of miniature, high-performance tracking units and receiving antennas in space should make global monitoring of animal migration possible. Really, ICARUS exemplifies both the direct progression of classic telemetry and its radical reversal: "The idea behind ICARUS arose out of discussions with the inventors of wildlife telemetry, William Cochran and George Swenson of Illinois. It had become obvious that in the field of ecology, information about individuals was the most fascinating thing out there, but that there was still no way to observe animals in the long term. For us, the springboard was Very Large Array, the astronomical radio observatory in New Mexico, which monitors small radio sources in the universe. George and I were sitting on some steps in the city of Gamboa in Panama, chatting about animal observation, and came up with the idea to launch a reverse radio telescope that would allow us to look at Earth and observe the small radio sources there. Thus, ICARUS was born."

The naturalists of today are inexhaustible multitaskers. They tinker with the smallest of the small and the largest of the large. They know as much about miniaturizing tracking devices as they do about managing satellites. Wikelski has a completely new communication structure in mind for the way in which we will approach nature in the future. The challenges are not just those inherent to the thing itself, but also come from within the professional field: "There were many problems. George Swenson had told me it would take about fifteen years to get the system off the ground. We have put fourteen years into it, at this point. I didn't believe it back then. The biggest problem was and remains the fact that we have to convince our colleagues—biologists and ecologists—that we really will have a breakthrough with these newest technologies. I am always hearing about how much better it would be to use the existent methods more extensively, rather than inventing disruptive, even crazy, new methods."

Things will always be a little crazy and disruptive in the midst of a paradigm shift, though. When vantage points fundamentally change and perspectives are disrupted, splintering can be expected. Wikelski is a disruptive thinker who has reversed the traditional lens, never once losing sight of nature in the process. Branches of research that have long worked intensively with technology, like astrophysics, are less fearful of the movement: both the European Space Agency and the German Space Administration support planning and implementation phases. What distinguishes the naturalists of today is their ability to think about the big picture. They no longer bury themselves in isolated taxonomical questions about individual animal families; instead, they have a bird's-eye view, as it were, of the world and its biospheres. This is yet another way in which they are worthy heirs of Alexander von Humboldt's cosmic conception. The aim is no longer to explore and protect tiny habitats from encroaching civilization; rather, scientists are now interested in a holistic vision of nature. The focus of their efforts is to harmonize nature and society, to find means and ways, and to solve the systemic problem of nature that arises when human society and nature are framed as two antagonistic systems. Wikelski is all too aware of the estrangement that set in when ecological movements began to separate humans from nature, thereby stylizing nature as something sacrosanct, a romantic image that over time faded into a memory of itself. At the same time, he is as well aware of the need for continued conventional conservation practices, in order to protect threatened habitats from the destructive dynamic of a global economy that sees nature as nothing more than a resource for the taking. Naturalists of the twenty-first century are tasked with mediating between the poles of an overly romanticized vision of nature, on one end, and the destructive force of modernity, on the other, and to search for a balance between the two. Wikelski knows that he must explore new avenues, and that he will have to resort to unconventional, even experimental methods. And he is convinced that problems will sometimes reveal their own solutions.

Extreme specialization and studied holism are thus connected in a new academic discipline that is positioned to serve the public, and that bears wide-ranging resemblance to von Humboldt's vision of "networked research." Humboldt maintained a carefully curated network of contacts that provided for a continual exchange of information. He was always active in a variety of disciplines, and he often sought out the advice of experts. He encouraged young scientists and incorporated their ideas in his thinking. He was successful in winning over politicians, receiving funding for his expeditions and gaining the support of heads of state. Viewed as a whole, these many qualities join to form the image of an interdisciplinary, networked approach that we might expect to see today. It has become clear that the complexity of modern nature can be understood only through multiple perspectives. Networked research poses surprising questions that constantly break knowledge barriers and provide insight, if disruptively: How does a young cuckoo find Africa without its parents' help? Do the global flu viruses that water birds carry to Europe originate in a southeast Asian primordial soup? How do the grazing behaviors of alpine cattle impact the diversity of alpine flora? How do baby turtles survive their first weeks deep in the ocean? Where, when, and why are animals confronted by problems they cannot master and that kill them? What do the blackbird's migratory behaviors say about global climate change? How could the study of the red-billed quelea's migratory behaviors be used to prevent famine in the Sahel?

Since an objective "nature" has never existed, and it has, instead, always been a conception conceived of by humans, the digital naturalists of today could also be called the (re)discoverers of nature. Martin Wikelski is working on determining what role this new nature—one that he has helped found, and that is in conversation with us—will play. He is cognizant of the possible problems and dangers that all human intervention invites, however well intended. But he approves of this transformation, in that he describes our current state of interaction with nature as stunted and impoverished. "In the future, humankind will return to the close exchange between humans and animals that defined the advanced civilizations of the past. I believe ICARUS and the whole idea of an electronic connection to individual animals will play a big part in this, and the Animal Internet will fundamentally alter humans' global awareness. We cannot yet even foresee the uses of distributed sensor networks of highly intelligent nodes, representing individual animals. There is going to be a revolution in the understanding of life on Earth."

The revolution starts at Martin Wikelski's front door. He is a cat lover, so what could be more convenient than to try out his tech inventions on his own pet? "Some of the first animals we tagged were our own house cats. I have always had cats and have one now, and every now and then, to experiment, I will put a tracker on her—one of the high-performance ones, at that. It was our goal, after all, to communicate with them, or rather, to let them tell us where they wander off to at night and just what, exactly, they do in the neighbor's yard."

... AND WHY "PROBLEM BEAR" BRUNO MIGHT STILL BE ALIVE TODAY

ON NEW FORMS OF COEXISTENCE

An important practical objective of Martin Wikelski's work is to lay the groundwork for the smart management of human-animal coexistence. News media regularly confer animals with fleeting celebrity status, putting this coexistence to the test through spectacle and keeping entire areas on alert for extended periods of time. One need look no further than at Bear JJ1, also known as "Problem Bear" Bruno, who caused a huge stir when he showed up and wreaked havoc in Bavaria in 2006, or at the coyotes that are increasingly spotted in urban areas in the United States, such as Manhattan's West Side. Conflicts between wild animals and farming communities occur all over the world: conflicts of interest that usually end unfavorably for the animals.

The problem of efficient wildlife conservation is usually not nearly as complicated as it seems. We do not often have more than a general idea of which human activities are responsible for animals' not surviving. After all, we do not have access to the central occurrences in the animals' lives.

The survival issues can almost always be traced back to negative environmental influences or habitat destruction. It has proven tremendously difficult, if not impossible, to use traditional methods of observation and analysis to determine the precise causes of death for many migratory animals, particularly birds, a class that is disproportionately affected by extinction. These creatures travel enormous distances, and a lot can happen along the way. Were we to follow them systematically via satellite transmitter during their migration, we could find out where the deadly dangers lie.

The North American Swainson's hawk (*Buteo swainsoni*), also known as the grasshopper or locust hawk, provides a telling example of the way in which tracking is already helping certain species survive. These birds spend the spring and summer in North America and migrate to South America in the fall. In doing so, they travel a distance of over six thousand miles. In 1994, biologists noticed a sharp decline in Swainson's hawk numbers in North Carolina. In an effort to understand the reasons for this decline, several birds were equipped with trackers, that they might be monitored all the way to their wintering grounds in Argentina. Upon arrival there, scientists discovered an area with over seven hundred dead hawks. The researchers spoke with farmers there and learned that the birds had probably died upon ingesting insecticides that had been sprayed on the fields, because Swainson's hawks love bugs. These realizations led to a discussion about regulating insecticide use in the region. In order to identify the relevant regions of Argentina, the researchers dug up data on Swainson's hawk migration routes that a research team led by Michael Kochert had gathered through the systematic tagging of fifty-four birds. The data were evaluated, visualized, and implemented as a course of action for concrete conservation measures.

Humans need to know where animals are, how they live, and what difficulties they are confronted with; only then can we be of any help to them. There are many other examples that illustrate how a transparent nature can help animals survive. The Cross River gorilla (*Gorilla gorilla diehli*) lives in the impenetrable rainforest between Nigeria and Cameroon. Very few members of this species are left. An estimated 250 live in the wild, having retreated to inaccessible mountain regions out of fear of humans. Since rangers and conservationists cannot easily follow them to their habitats, almost no data exist about these animals; only recently have researchers managed to capture any footage of the gorillas in the wild. The North Carolina Zoo and the Wildlife Conservation Society are therefore employing GPS trackers to study population distribution throughout the area and how the animals interact with the increasing human activity in the jungle. The result is systematic data and maps that can be used to help regulate habitat protection measures. It allows lines to be defined for guiding the coexistence of humans and these vulnerable species.

Peter Walsh from the University of Cambridge takes it a step further. He is thoroughly convinced that ecotourism targeted at large primates would markedly improve their chances of survival. His actions clash with the zoological credo stating it is best for the gorillas to remain as isolated from humans as possible—for reasons including the fear that diseases could be transmitted from humans to the animals. In light of current threats, however, it seems that costly ecotourism, a popular activity for the very well-heeled, could be the only possibility for saving the animals, and that the human contact it entails is a calculable risk worth taking. Data published by the International Union for Conservation of Nature (IUCN) show that over the next three generations, around 80 percent of Western lowland gorillas will die as a result of pandemics, poaching, and habitat destruction. Money spent by tourists, who could be directed straight to the GPS-tagged animals, could be invested in regional infrastructures and create new jobs. Taxes would pour in, thereby increasing the likelihood that the state would support gorilla conservation efforts. This targeted interaction of humans and animals seems instinctively to contradict with images of Eden, of a nature that can take care of itself and that may not be touched. According to Peter Walsh, however, this idea subscribes to a dangerous myth that itself turns against the animals: "Nature has always been man-dominated, and if we don't get a little more invasive, it's not going to remain Eden."

It would appear, then, that the fundamental problem with so many nature conservation efforts is this traditional understanding of "wilderness." To this day, it seems that the image of an intact natural world must include wild animals that exist totally independently of humans. But it isn't that simple anymore. The IUCN's regularly updated guidelines amount to the manifesto for nature conservancy. The IUCN, one of the most powerful nongovernmental organizations for nature conservation and the architect of the "Red List" of threatened species, explicitly presents a new and expanded understanding of "nature" in its most current guidelines. In the past, the guidelines were built around an understanding of nature that placed strict boundaries between humans and animals and employed all available means in maintaining them. For a long time, for instance, animals were reintroduced to the wild by being simply turned out and left unattended. That doesn't work anymore. Reintroduced animals like bearded vultures or wild horses require intensive support in order to survive. The great outdoors are now dotted with arterial roads, ski areas, and biking trails. This is why animals are being tagged and tracked: in order to be protected from humans, on the one hand, and on the other, to expose humans to their complex and hitherto hidden lives, thereby prompting changes in behavior. Wilderness 2.0 signifies a natural world that speaks for itself and that could be open to the public by means of targeted tourism programs.

Geopolitical shifts can also be problematic for animal habitats, because whenever political systems crumble and new borders are drawn, animals are faced with danger. Poaching tends to increase in the midst of political chaos, and newly created borders present new challenges, especially to the more nomadic species. An extreme example of this is the saiga antelope (*Saiga tatarica*). These animals live in the Eurasian Steppe, which stretches over 3,700 miles from Russia to Mongolia. There are no other large mammals on Earth that have been decimated as quickly as this member of the gazelle family. Around one million saiga antelope still existed in the 1990s. Ninety-five percent of them were killed in the first ten years following the collapse of communism, primarily through poaching, which had been kept under better control during strict Soviet rule than in the post-Soviet transitional chaos. Today the numbers have risen back to one hundred and fifty thousand, spread over many areas. Aline Kühl-Stenzel coordinates the saiga protection project at the United Nations Environment Programme. Animal conservation, Kühl-Stenzel argues, must be approached from two sides. For one, something must be done to elevate the social and cultural conditions of hunters and poachers to provide them with alternative sources of income. At the same time, the saiga antelope are receiving protection from poaching by the transparency gained through their tagging. Furthermore, determining the exact migration routes should also help establish ecological corridors in a geopolitically complex region, allowing the animals to reach their birthing grounds. It is now known that antelope travel upward of eighteen hundred miles each year. After a year of tracking saiga, the areas where animals go during the mating season have been identified and turned into nature preserves. Finally, knowing the animals' movement patterns can shed light on how the ecosystems of the future will shift.

PROBLEM ELEPHANTS, PROBLEM BEARS, PROBLEM WOLVES

Tracking wild animals also helps promote coexistence of potentially dangerous creatures and humans, which has become increasingly difficult as habitats shrink. The Animal Internet can be used to monitor specific "problem animals" that move throughout a wide radius, and help them coexist with humans in a world that has been splintered by human activity and highways. In southern Nepal, a GPS device was used to track and treat an injured tiger that had broken into a vacation resort. The tiger was later released back into the wild at a location farther west. The GPS data can now be used to see the routes the tiger is taking, and potentially to prevent it from wandering back into human territory. In the past, the tiger would simply have been killed.

In Kenya, farmers and elephants share a living environment. The animals should have decent living conditions, and the farmers work hard to secure their meager livelihood. This conflict very clearly gives rise to a moral dilemma. Whose side to choose? The farmers' or the threatened elephants'? A survey conducted today in any large European city would almost certainly find most people taking the elephants' side, because our decisions are steered by a collective guilty conscience about nature, and not by social consciousness. Technology can help solve this moral dilemma.

For the last ten years already, elephants in Kenya have been tagged with GPS devices. The data on elephant locations and migratory patterns have been used to obviate the conflicts that arise when elephants destroy crop fields. The data on migration have prompted many of these routes to be turned into protected corridors. At one such location, in fact, a highway underpass was constructed. Furthermore, particular problem elephants that are known for running riot in these impoverished farmers' fields are equipped with special units that—should one of these tuskers venture too close to a farm or village—will send a text message to the rangers, who can then quickly locate and redirect the animals. Elephants are so intelligent that they take note of these rebukes—or interactions, when it really comes down to it—and avoid these virtual fences in the future.

Zoologist Josef Reichholf has identified another important possible use for digital technologies in human-animal relationships. These can be grouped under the umbrella term of "smart farming": "One of my greatest hopes is that the transparency of food production processes, which has increased via social media, will lead to improved regulations. Because I need to experience it for myself, to see what I'm living on and where these products come from; what the conditions are in which they are produced; and how the animals producing them live—only then can we react accordingly to the mass production of cheap meats and raise awareness in the general public. Because politicians won't be the ones to push through these changes. That's the lesson we've learned from decades of so-called agricultural policy."

For a long time now, the problem of human-animal coexistence has not been a problem facing populations in the Serengeti or the rainforest alone. Europe's woodlands are seeing the return of raptors, joined by lynxes, bears, and wolves. The tale of the brown bear Bruno—who wandered from Italy to Bavaria and caused a stir in the foothills of the Bavarian Alps before meeting his tragic end in a hail of bullets—is a good example of irresponsible wildlife management and bad public relations. Bruno was the first known bear in Germany in 170 years. He appeared, as yet unnamed, on May 10, 2006, in the Austrian town of Galgenul. He had just woken from hibernation and quieted his hunger with six sheep. Over the course of his travels, he always stayed near villages, but remained unseen. People saw only what he had done: emptied henhouses and pigsties, bloody carcasses. Then, on May 19, he crossed from Austria into Germany. He was identified as JJ1, the first-born of an Italian bear named Jurka. The Austrians had tried to chase JJ1 out of residential areas using rubber bullets and firecrackers, to no avail. The Bavarian state government declared an open season to shoot the bear. At this point, JJ1 became known as Bruno. Finnish bear hunters flew down to catch him alive. Bruno lay low and kept out of sight. He never returned to the scenes of his crimes, his feeding areas. The bear gradually became a celebrity, even though hardly any pictures, let alone video footage, existed of him. Only a handful of hikers and cyclists managed to upload a few snapshots of Bruno to the Web. Finally, Bruno was shot, unleashing heated discussions in the media. The name of the hunter who delivered the fatal shot remained secret. Photos of the dead bear were not published.

The fight regarding Bruno, between the authorities on one side, and animal advocates and the public on the other, was a fight about pictures. The animal's invisibility within civilization scared people, and those responsible used this to their advantage. What would have happened if Bruno had been outfitted with a GPS tracker and a GoPro camera back in Italy? After all, he had already earned himself a reputation there for his bullheadedness in entering developed areas. Had Bruno had a tracker, had his movements been broadcast in a video stream, then he would probably not have become a "problem bear." He would probably have been caught right at the border of the Italian national park. Had he had his own website, where people could track his route, he would not have become the sinister lurking threat he was portrayed as in the tabloids; instead, the public may even have cheered him on, identified with him, donated money to support wildlife conservation. If nothing else, this form of transparency would have prevented the cloak-and-dagger hunting operations. Josef Reichholf has a strong opinion about this: "If that bear had had video monitoring, it would probably still be alive." He said the same applied to other creatures that have been sighted in urban environments, such as coyotes in various New York City parks or a boar wandering through Berlin with her piglets. "It doesn't just apply to the lion in the Kalahari or the jaguar in the Pantanal in Brazil anymore."

The Internet is bringing humans and animals closer together. This new form of coexistence is an absolute necessity, particularly in cases of predators in densely populated areas. Even the wolves have returned to Germany's forests. Not everyone welcomes the return of these creatures, which remain steeped in legend. Sheep farmers and hunters are not the only loud voices in the debate. In the rest of the population, preconceptions that had been thought debunked for the last hundred years are undergoing a renaissance. In the areas that have seen the return of wolves, parents fear for their children's safety on the way to school: shades of Little Red Riding Hood.

Markus Bathen is the wolf commissioner for NABU (NAturschutzBUnd), Germany's Nature and Biodiversity Conservation Union. A wolf commissioner is something along the lines of a ranger in a U.S. national park. NABU is an NGO, founded in 1989, comparable to the Sierra Club in America. It takes on the task of establishing and maintaining nature preserves and works to educate society on environmental topics. One of the many tasks NABU has taken on is shaping the conversation surrounding wolves' controversial return to the German cultural landscape. NABU agents are most active in Lusatia, a region between Germany and Poland now home again to multiple wolf packs, in some places living close by to towns and villages. There are three established natural parks in Lusatia. Following their near extinction, wolves returned from western Poland to Germany of their own accord. They can generally hold their own, and Markus Bathen does not need to fight for their immediate survival. To create a lasting impression, however, it will be necessary to dismantle the negative images that people—especially those living in close proximity to wolves—still have about these animals. Bathen's motto is "facts, not fairy tales." In addition to slideshows, school workshops, and info booths, Bathen has also turned to social media to familiarize the German public with the Big Bad Wolf. There is a lot he could learn from the rangers in Arizona's Maricopa County, who are pioneers in the use of digital communication in environmental education. Bathen describes his starting point as follows: "Knowledge of wolves is minimal, making the fear of them that much greater. Wolves are nocturnal. You don't see them. 364 days out of the year, they are absolutely not present. Then, on day 365, they leave some bloody remains, and it comes as a total shock! When people can follow wolves' current activities via social media, then they can share in these animals' everyday lives, and the problematic cases can be better explained and catalogued."

WHY TECHNOLOGY IS NOT ALL BAD, AND NATURE NOT ALL GOOD

DATA PROTECTION FOR ANIMALS AND THE POSITIVE SIDES OF TRANSPARENCY

The time has come to consider a fundamental objection to the Animal Internet: viewed with a critical eye, digitizing nature effectively amounts to subjugating it to technology. Do we really want to turn nature into a single, massive laboratory for unrestricted human experimentation? In doing so, are we not destroying the one crucial source of creative power that can help us out of the vice grip of civilization? The conclusion we can reach is that through quantification, the Animal Internet is destroying the quality of what nature means to the world.

Following the Internet of Things (IoT), the Animal Internet is now completely wrapping up the virtualization of the world. Until now, nature has been the one area that the Internet could not touch, the last remaining retreat from ubiquitous digitalization. It's the last data-free space. When I've had enough of the Internet, I can turn off my devices and wander through the woods. To recharge my (nonelectronic) batteries. To clear my mind. To escape reality. This green immediacy and the energy it radiates is destroyed the moment I take out my smartphone to check an app created to help me trace a woodcock or a pine marten. Are we in danger of losing the ability to *see*? And isn't that the very crux of the study of nature? "It's not what you look at that matters, but what you see," the natural philosopher Henry David Thoreau wrote. Wasn't he right?

When we try to rationalize and articulate our unmistakably critical gut feeling about the Animal Internet, there are three concrete charges against the digitization of the animal world that we must examine.

First: The Internet is unnatural. As a form of technology, it does not belong in nature. When we occupy ourselves with the Internet, we are occupying ourselves with the opposite of nature: a machine.

Second: In combination with the Internet of Things, the Animal Internet poses the threat of total digitization. Humans become prisoners of their own inventions, slaves to the devices running the world.

Third: The new transparency of nature puts animals at considerable risk. Animals we can track are animals that can no longer hide from us. Rather than allowing everybody to pinpoint the locations of various endangered animals, we need to suppress that information so the creatures in question can be protected.

The line of argumentation reframing the Internet as the new solution to nature's problems sounds like a lazy compromise at first. We must question how the Internet, the least sensory and hands-on of all media—in other words, the most purely virtual—could ever be considered close to nature. At first glance, the Net seems to be the antithesis of organic existence. Dig a little deeper, however, and we hit upon an interesting constellation, upon analogies that are worth analyzing, that reveal the Net to have thoroughly organic structural characteristics. The Net itself is a "natural" construction. Google can be seen as a sort of embryo that is forever developing and taking on new forms. The Net is not a finished product; instead, its very essence consists in furcated growth, in meandering. The Net also exhibits many qualities that the scientist Edward O. Wilson calls biophilic in form since the Net is in a state of constant permutation. It is a continually self-reorganizing living system and it is—like nature—complicated and fickle. The Net divides into branches, it is organically arranged, it reflects the way nature is organized, and finally, it is an attempt to master the complexity of information and resources: a task that every naturally occurring system must take on, if it wants to survive. The Web could be described as a "natural" system in all of these regards.

And that's exactly why we love it. The Internet is anchored in our biological history. Humans created it out of a love of life and to imitate nature. For this reason, the Net is the perfect medium for the new animal dialogue. It is a synthetic continuation of nature and part of the biophilic revolution.

On the other hand, the Net makes us extremely biophobic. It distracts us from direct contact with the living world. It encourages withdrawal into an electronic existence, distances us from the world, and leads to vicarious engagement. The Net and electronic entertainment industries are giving rise to new disorders. The challenge is thus to incorporate the Net and digital technologies—networked transmitters, tags, websites, blogs, social media outlets, smartphone apps, wildlife cams, and so on—into an expanded concept of biophilia: in an understanding of ecology that moves beyond the separation of nature and technology, wilderness and development, in favor of an integrative perspective. It is a new nexus of technology and nature that does not readily reveal itself to us, because it is buried under layers of cultural textures.

Given that many people are already overwhelmed by smart refrigerators or running shoes, though, how are we meant to master the Internet of Nature or integrate it into our daily lives as a useful structure? To answer this question, it's worth looking back at the history of the Internet. On the one hand, tagging animals and gathering digital data on their movements is part of a new era in the history of nature studies that will see an exponential growth in our knowledge of animal life, shining a light in the darkness. On the other hand, however, this Net also marks the beginning of a new chapter in the history of the Internet. A third Internet has been created.

The history of the Internet is often depicted today as a purely technological development. It appears as a succession of digital milestones, characterized by higher bandwidths, more extensive interconnectedness, and improvements to mobile devices. In a way, these technological achievements simply mark transformations occurring in the inner life of the Internet, and not its actual history. The real history of the Net should be seen less as a series of infrastructure models, and more as a story of the fundamental changes and revolutions the Internet unleashed. Every phase of the Internet changed the world in a very specific way. Every phase of the Net brought about a paradigm shift.

To date, there are three discrete phases of the Internet—and three corresponding shifts in societal reality—that we can identify. The first Internet is the Human Internet. It connects individuals and groups, both professionally and personally. It has changed society through new forms of communication, interaction and participation, exchange and access to information. To a certain extent, this first Internet has changed the way we define friendship, relationships, reputation, and participation in social life. Social media is primarily responsible for the transformative power of the first Internet. For people condemned to silence through authoritarianism, these communication channels have provided an opportunity to speak, contributing in some cases to the fall of political systems and the creation of new societal realities. The Internet has thereby shown itself to be a powerful vehicle for political organizing. It compensates for the disadvantage grassroots opposition movements have traditionally faced, namely modes of communication inferior to those of the government. By giving the masses a voice, it has enabled or intensified instances of social upheaval.

The second Internet is the so-called Internet of Things (IoT), which gives everyday inanimate objects a voice. Objects in this Internet realm are equipped with electronic elements that provide these things with "intelligence." For the time being, the second Internet has changed our everyday life only by making it possible to track objects in motion (like parcels or other delivered goods). Increasingly, though, the world of things is being penetrated by electronic structures, that is, structures that perceive and communicate with their surroundings. Everyday objects are thus transformed from lifeless, limited assemblages of material to humans' intelligent, verbal partners—for example, a pair of running shoes with embedded chips will tell us how many miles we just ran, and how fast, and how many calories we burned, or our T-shirt might let us know when we're sweating a little more than we should be.

One could certainly wonder whether we need all this. One could ask why things should learn to think and speak, and if it wouldn't be better if they were simply "there" for humans—things to use or watch—rather than being given a voice. The technological elevation of objects in some ways makes them independent subjects. This carries with it the danger that human freedom will become increasingly limited. There is the risk that we will become slaves to technology and relinquish our privacy. This would be the case if things truly started telling us what to do next, if our free will started to be strongly influenced or even dominated by our things' data.

The problematic dynamic between humans and machines in the age of the IoT could also be judged differently, however. This technology also has the potential to liberate us from the dictatorial clutches of our things. The process of industrialization showed us that humans can become slaves to machines when subordinated to them, like Charlie Chaplin in *Modern Times*. If we do not want machines to carve away human rights, then we need to enable them and other things to understand what we want.

We must try to elevate them to our level of communication. Through technology, we need to get them to speak our language and answer us. After all, a stupid machine is undeniably a greater tyrant than an intelligent one. We can program an intelligent machine to align itself with the demands of our lives. It may still be awhile until we find it useful to have a refrigerator that pings our smartphone when we're out of milk, telling us to pick some up on the way home. There will also be people who prefer to go without their morning latté than to have a machine dictate their shopping list to them. However, a refrigerator that knows what we want to hear, and when, and with what urgency, is definitely less tyrannical than a refrigerator that keeps quiet until it gives up the ghost one day. The IoT is therefore not establishing a new structure of slavery, but a structure of autonomy, provided we know what we want to accomplish with it.

Following the second Internet, that of things, is the third generation of the Net. It is the Animal Internet or, in a broader sense, the Internet of Nature. It gives animals a voice. Just as the Human Internet has affected society, and the Internet of Things everyday life, the Animal Internet is changing a fundamental aspect of life on Earth—it is changing the image we have of nature. The new image of nature is no longer idealized, it's realistic. It confronts us with the concrete interactions between animals and the problematic environments in which they live. In doing so, the Animal Internet formulates three new basic animal rights that have the potential to revolutionize the human-animal relationship. These rights, I would argue, are as follows:

First: Every individual animal has the right to an identity. *Second:* Every individual animal has the right to be known and protected by humans. *Third:* Every individual animal has the right to ideal living conditions in its respective habitat.

Based on these three demands alone, it is clear that the Animal Internet is not anthropocentrically oriented—that it is not structured around humans. We are not looking at a modification of the Human Internet that expands to include animals; rather, the *subjects* of the Animal Internet are *the animals themselves*, as unique individuals. Once the Animal Internet has gathered representative amounts of animal data, nature will no longer be the way we imagine it, or the way we wish it to be, or the way we have had it explained to us. We will start to see it as it truly is at the moment of viewing, and we will also see how individual animals have adapted to concrete situations for which we are responsible. The Animal Internet is therefore radically objective and radically subjective at the same time: on the one hand, it objectifies our image of nature and lays new groundwork for scientific and ecological discussions, in that it collects raw data that translate into new knowledge. On the other hand, it turns animals, which we are used to thinking of as objects, into subjects with their own back-story and fate. It is not the rights of a species or a genus that emerge from the Animal Internet, but the personal rights of the individual animal subject.

The third Internet is reminiscent of the existential dual-ism that so closely connected humans and animals in the nineteenth century, that golden age of work animals and the deep-rooted coexistence of man and beast. Because what, exactly, does the statement mean, "Every individual ani-mal has the right to an identity"? It means nothing other than that a dialogue must arise between humans and animals endowing the latter with an identity. This means that humans are now required to *grant* wild animals an identity. Work animals are "identifiable" by the use they serve for humans. These are not species in need of protection. Allowing animals to have individual personalities is the defining difference between human and animal ethics. Every animal has a backstory and a personality, even if it does not have a chip. It simply cannot *tell* this story. It cannot share it with humans. The animal does not have a personality until humans are prepared to grant it one. The same applies to house pets. We are comfortable attributing a personal destiny to our dogs and cats, because they live in our social sphere, because they are part of our social framework. Wild animals live outside of this social zone, which is why they do not have a personal identity in our eyes; at most, we see a set of characteristics applied to the species in general. The Animal Internet will change this, thereby laying the foundation for new animal ethics. Interconnection to the human world gives animals the opportunity to articulate their personal histories and to share their identities with humans. In this way, the data stream constitutes their right to exist.

This leads us to the third argument that can be made against the digitization of nature: Do we need data protection for animals? And what are the basic features of the ethics of the Animal Internet? There is a concrete example that illustrates the move from the Internet of Things to the Animal Internet, and especially the important moral shift from the rights of a genus or a species to those of individual animals. Let us compare the talking fridge—which serves repeatedly as the paradigm of the IoT—with a "talking" wild bird, such as a waldrapp, that has been tagged for the purpose of tracking its migration route and that updates its current location continuously via Facebook. The two cases are obviously fundamentally different. The first is an inanimate object without rights, the second, a living creature with vested rights. The refrigerator is not a moral subject, but the bird is, to a certain extent. The very question of whether and under what circumstances it is justifiable to tag an animal distinguishes these two cases from each other. After all, this act requires the capture and sometimes the sedation of the animal; these are actions that cause it stress. The ethics of the Animal Internet, which are needed to define limits to this interference, must answer the following questions: Who may interfere with animal life, under what circumstances, and to what end? When is such an intervention unjustifiable? How many animals must be tagged in order to provide representative information about a species? What happens to animals that are not tagged? Do these animals have no rights, simply because they have no voice?

For all their differences, both cases bewilder us with an analogy that revolves around the terms "responsibility" and "transparency." When we "make objects talk," we're doing more than just fiddling around with a freaky idea. We are shaping our relationship with the inanimate world, moving in the direction of a deliberate, responsible coexistence of humans and machines involving communication between intelligent life and material objects. It's no different with the Animal Internet: When we tag animals, thereby providing them with a certain power of speech, we are fundamentally reshaping our relationship with nature, setting ourselves up to read the animals' identities. We consequently assume responsibility for animals we were not aware existed till recently. Transparency in nature engenders awareness, which itself invites obligation.

Do we truly need data protection for animals, despite this nascent matrix of responsibility? It is, of course, a prime concern that the gathered animal data be transmitted and stored securely. There's not yet a legal framework for the large-scale collection and storage of animal data. After all, animals are not governed by human law, they are not individuals. Instead, they are merely recognized as representatives of their species. While they have the right to treatment or living conditions appropriate to their species, they do not have the right to have their personal situation or concrete living conditions considered on an individual basis. But their legal status could already be changing. The New York State Supreme Court has recently considered arguments that chimpanzees should be granted a writ of habeus corpus, which has been used to prevent people from being unlawfully imprisoned. The writ, though it was ultimately denied, would have allowed chimpanzees to be rescued from living conditions that are—somehow, though by whom?—deemed unjust.

From the legal perspective, the Animal Internet presents new challenges. Beyond the legal dimension of the topic, however, the general question remains as to whether it is fundamentally good or bad for people to know where animals are. This applies to all animals, of course, but for highly endangered and coveted species like apes or big cats, it can take on dramatic magnitude. Should these animals' positioning data fall into the hands of poachers or tour organizers, for example, the transparency intended to protect could very quickly turn against them. It's still too early to draw a general conclusion from existing evidence on this topic. Examples have indicated, however, that transparency tends to protect animals from poachers, rather than having the opposite effect. In the case of the waldrapp, the ability to follow the birds' migration to Italy has resulted in the Italian hunters' no longer going after them. The birds are on the collective radar now, and that makes them too risky a target. The Internet has saved the waldrapp. This is no exaggeration. There are few animals that have ever needed as much compassion and technological aid as these archaic ibises. Zoologist Johannes Fritz, who leads the European Waldrapp Project, points out the central task of reintroducing these birds to the wild, as well as to the general public. According to Fritz, the waldrapp "is an animal that has long disappeared from popular consciousness." Technology is making the bird visible again, reintroducing it to us. The trackers make it possible to create an audience for the bird. At the same time, Facebook is becoming a tool to combat poaching in real time. The more transparent the birds' movements become, the more people "friend" and chat with the four black fowl on Facebook, the greater the pressure will be on the hunting associations and authorities. This is exactly why social media engagement with these baldheaded ibises has gradually been expanded. The animals can now be followed using the Animal Tracker smartphone app. The software shows their position: "With a single click on a particular animal, you can see its entire life story," raves waldrapp expert Fritz. "The animals are becoming real personalities with whom we can communicate, who become our friends—and this is good for nature." And for humans, one might add. And especially for me. Because I, too, am friends with a 'rapp. His name is "Shorty."

WHY ANIMALS WERE ALWAYS FRIENDS OF HUMANS

A LITTLE STORY OF EMPATHY

Shorty the waldrapp is a Facebook friend of mine. He's a strange bird. A little unconventional, but not unlikable. He looks much older in his profile picture than he actually is. He's bald, his skin wrinkly and furrowed. However, he gives off a tough, tenacious, and thoroughly youthful vibe. He wears his thin neck feathers in a crazy plume. He spends winter in the south, in Tuscany, for health reasons.

Shorty doesn't yet live independently in the wild; rather, he's part of a reintroduction program. He was raised by humans and is now learning what life in the wild is all about, because Shorty has forgotten what it means to fly over the Alps in search of warmer climes, come autumn. His human foster parents have the arduous task of teaching him how to migrate. They accompany his first migration in an ultralight aircraft, in the hopes that this action will activate a genetic memory in the ancient bird species that will allow him to guide others of his kind over the mountains next autumn. In order not to lose him, he has been tagged with a GPS tracker that sends signals online via satellite. This allows people to see where Shorty is at all times. The rest of the waldrapp flock are also equipped with tracking units. The data are made public on a Facebook page (www.facebook.com/Waldrappteam). Almost every day, visitors can view maps for a detailed account of the waldrapps' current location, how they are doing, and what they may recently have encountered. The interaction between humans and animals on the Facebook page is intense. Waldrapp friends search for the animals, photograph them, and upload the images. However incomplete, a multifaceted picture of these animals' everyday lives thus emerges.

There are many other ways in which social media could be used for communication between humans and animals. Taylor Chapple is responsible for tagging great white sharks in the North Pacific. He works at the Hopkins Marine Station of Stanford University. Shark Net, the attractive iPhone application he and his team have developed, allows users to follow individual great whites. The animals do not yet have a Facebook page or blog, though. "It's coming," Taylor stresses. "It's in the pipeline." The goal for the great white, an endangered species, is also to dismantle preconceptions and familiarize people with the sharks' way of life. Social media can help bring the sharks closer to humans. Until the site has been launched, however, the European Waldrapp Project remains the benchmark for demonstrating how digital communication with animals can work, as well as its natural limitations.

Shorty is my first animal Facebook friend. When he started on his way from the breeding colony in Burghausen, east of Munich, to his Italian wintering grounds last fall, I was eager to see how it would feel: to be in direct contact with a wild animal, to soar over the Alps with this bird, peering over his shoulder like the Swedish fairy tale character Nils Holgersson on the backs of the wild geese. The very notion of "following" a wild animal, in the truest sense of the word, staying hot on its track, promised totally new experiences and insights that far surpassed anything out of a documentary. It was the real-time allure that drew me in.

The second time I visited the page to check on Shorty's current status, however, some doubts crossed my mind: Wild animals as Facebook friends? What has the world come to? An increased closeness to nature is desirable, but does this return to nature have to be digital? Isn't it better to go into the woods and forage for mushrooms? Isn't the Net more a part of the problem than a part of the solution?

As soon as Shorty struck out for his wintering grounds, however, the doubts lifted and things got interesting. It didn't take long for the emotional tie to the animal to form. I was sharing in the thrill of things. How many miles will Shorty manage today? Will he reach his destination? Will he find the right way? That sounds easier than it is. Shorty actually did lose his way. He landed in Switzerland, in an area that is too cold and desolate, even in late autumn, for a waldrapp to spend its winter. Along with many others in Shorty's network of friends, I immediately wanted to know: Does the bird stand any realistic chance of survival down there in the ice, snow, and cold? Even those in charge of the waldrapp project doubted the site's suitability as a winter spot. Johannes Fritz, the project director, wrote the following concerned Facebook post on January 18:

No word from Shorty for several days now; have any of our Swiss friends seen him recently? It would appear as though the weather has changed in Switzerland, now, too.

Three days later, and still no sign of him:

There have been no more reported sightings of Shorty since last Wednesday. It would appear as though he's flown on, perhaps even just closer to Lake Zug. Based on past experience, I'm optimistic that Shorty will manage in the current weather conditions, especially because he clearly found enough food before the weather changed. In consultation with our team, staff members at the Goldau Landscape and Animal Park have been trying for some time to attract the bird with food, unfortunately without success. Still, many thanks to all involved! We will now wait till new sightings are reported, and in the meantime, we're thinking about how we might catch the bird, should he fail to respond to our attempts at luring him in.

Catching the waldrapp proved a difficult undertaking. On February 11, Johannes Fritz therefore posted a call to action intended for Shorty's Facebook friends in Switzerland:

In order to launch a new attempt at catching Shorty, the bird's location needs to be at least halfway predictable. We are therefore calling all people on Lake Zug who have the time and motivation to search for and observe the bird. It would be ideal to know where he sleeps, or perhaps a certain meadow where he spends his days. ... We will immediately post new sightings here on Facebook, in order to assist in the search for Shorty.

The appeal worked. Local Shorty fans supported the search for the waldrapp and uploaded their photos onto the Facebook page. The rundown on February 12 was as follows:

Herr Brunhold has some really interesting information on Shorty's whereabouts:

Saturday, 02/09/13, 16:00: Brief spells of sunshine and snowfall. Town of Risch, Zweiern area (south, near Freudenberg Palace). Shorty amongst the graylag geese.

Sunday, 02/10/13: Frequent sun, but cold. A passerby reports a Shorty sighting in Risch, near Dersbach Manor, north, near Freudenberg Palace.

Monday, 02/11/13, 15:15: Heavy cloud cover, occasional bursts of sunshine, around 0°C. Risch, Dersbach Manor, north, near Freudenberg Palace, in a sheep pen. Shorty pecking busily at food in the sheep pasture. Approx. 20 m distance from the sheep.

The waldrapp appears to have made his home on the western shore of Lake Zug, somewhere between the public beach in Hünenberg and the village of Buonas. This is, however, an area spanning about 2.5 km.

Herr Brunold's other photos show the bird eating. His condition still appears to be good.

It seemed impossible to catch Shorty without professional help. February 14:

> Our associates at the Max Planck Institute in Radolfzell have agreed to help catch Shorty using so-called cannon-nets. I think this will improve our chances of finally catching the bird, in order to unite him with others of his species in Tuscany. Because as well as Shorty is clearly managing in Switzerland, it is far from adequate for him. And when we remember the fact that he belongs to a group of about 25 migratory waldrapps, then our efforts are certainly justified in trying to ensure his optimal chances of survival.

Despite all attempts, the animal still evaded capture. There was no shortage of helpful suggestions from the increasingly active waldrapp community:

> Maybe they should try using "live bait" ... put his best pal in a live animal trap ... A dummy ... + play a recording of their "song" ... they use CDs to attract swifts and that works.

Criticism of the attempts at capture soon entered the discussion. Several animal lovers argued that the waldrapp should be left to its own devices. It's a wild animal, after all, that may need human support up to a certain point, but that must then learn to live on its own.

Shorty's odyssey, meanwhile, seemed to know no end. On April 3, Johannes Fritz posted a passing update:

Just a little new info on Shorty, reported from Switzerland. He was seen with the graylag geese yesterday, Tuesday, by Herr Simeonidis on Lake Zug, apparently still in good shape.

There will not be any new attempts at capture, as long as conditions do not unexpectedly change. We expect to see our Swiss bird back at his breeding grounds in Burghausen in the coming spring, and we are already looking forward to it.

In the meantime, I hope to continue receiving reports of sightings from our dedicated Swiss friends.

As the first gray-brown chicks start hatching in Burghausen in May, it became evident that Shorty had decided to stay in Switzerland for the time being. On May 8, we read:

There's been news from Shorty. Martin Brunhold visited him a few days ago and had this to report:

"After strong hailstorms in Risch on the evenings of May 1 and 2, I was gone for a few days and couldn't keep an eye out for Shorty until yesterday. I found him at 15:00 in a field near Buonas, in sunny weather. About an hour later, he was poking around a mown meadow near Zweiern, alone, and he was not bothered by the people out for their Sunday stroll. I could tell that many of these walk-

ers knew our waldrapp, based on their conversations and reactions."

Eventually, Shorty did return home to his flock. While they
soared, shining black, over the mountains, I sat at home
in front of my computer, watching the blue and red dots
moving across the map, and breathed in the air of freedom.

The wilderness is undergoing a revival in the living
room.

THE SIMILARITY OF THE OTHER

But how close is this digital closeness really? And what
does it say about the possibilities of a renewed interaction between humans and animals? The existential bond
between living beings is what first made animals attractive
to humans as companions. This half playful, half serious
closeness has gotten away from us—entirely without the
influence of the Internet. The linchpin of the new ecology, the springboard for revaluating the paradigm of green
ideology, is the awareness humans have of animals. Classic
green thought is built upon the idea that all animals behave
in a way that is typical of their species. It treats animals like
an abstract group, and views single members of this group
not as individuals, but as interchangeable representatives
of their species. An emotional tie cannot be formed with an
abstract image of a species. A symbolic representative will
never be an individual. The Animal Internet's opportunities
for social interaction allow for breaking through the logic
of abstraction. They allow humans to connect socially with
wild animals, as we experienced concretely with Shorty.

Only this can help create a new view of nature. "We should not," says zoologist Reichholf in this regard, "write off this kind of interactive connection as 'pathetic fallacy.' Many animals would benefit from emotional humanization ... [because] it's the closeness to other living beings that we're missing and that conservation policies are blocking in such a ridiculous manner."

The digital cosmology emerging in tandem with the Animal Internet relates to this weakened, but not totally extinguishable, human empathy for animals. In the past, animals were not merely practical helpers and useful partners in everyday human life. Animals are closely connected to the emergence of human cultures and civilizations. The first images humans painted on cave walls were of animals. The first paints humans used were probably animal blood. According to one theory, human language began as an imitation of animal sounds. This argument is made both by Plato, in *Cratylus*, and Rousseau, in his *Essay on the Origin of Languages*. Ancient mythology and poetry provide a living mirror of this symbiosis. They depict humans, gods, and animals in the flowing transitions of metamorphosis and metempsychosis, of morphogenesis, shape-shifting, and rebirth. Ovid's *Metamorphosis* contains many examples of this: Jupiter changes Lycaon into a wolf and Io into a cow, Diana transforms Actaeon into a stag, Athena recasts Arachne in the form of a spider, and so on.

Experiences with the animal kingdom allowed humans to take possession of the world, both practically and metaphysically, to understand physical connections, and to guess what was happening beyond the realm of the visible. Animals provided the structure for human awareness of the world around them. They gave it form and shape, they made the unknown their own. Animals were used as signs and symbols for defining the cosmos, making the infinite and incomprehensible vastness of the universe seem less alien, and bringing it all closer to humans. The wealth of animal forms predestined them for use in making the blanket of stars describable, in the first place. When humans looked at the heavens, they saw in them the creatures that surrounded them on Earth: eight of the twelve zodiac signs are animal-based. In Hindu cosmology, the earth is carried by elephants standing on the backs of turtles. People also used animals to reduce the threat of the unknowable by interpreting their behavior to prognosticate, again calling to mind the proverbial canary in a coal mine. Humans were as wont to read the entrails of sacrificed animals as they were to watch birds in flight for signs about the future.

It's easy for modern humans to write off this kind of divination as superstition or magic, and to consider themselves above the foolishness of ancient peoples. This supposed foolishness, however, exhibits an essential human trait, namely the tendency to describe the unknown— which is fundamentally threatening and potentially destructive, given our inability to plan for it—in terms of the known, thereby making it more tangible. Even today, humans are constantly pushing the boundaries of their emotional "safe spaces"—better known in today's jargon as "comfort zones"—and in doing so, they use all manner of traditional symbols, including animals. In a thousand years, humans may find our planting a flag on the moon to be as grotesque and irrational as we consider hepatoscopy, the liver examinations practiced by the Babylonians and Etruscans, to be an absurd method of prognostication and diagnosis. Upon seizing the moon, however, the act of erecting the flag was an important symbol, not only of victory over the Soviet competition, but of triumph over what the moon represents—the cold and the unknown. There's nothing different between this flag declaration and humans charting the night skies with the animals they hunted and that were part of everyday life.

For prehistoric humans, therefore, animals were not only of this world, but most decidedly of the great beyond, as well. Animals participated in both worlds. They were at once mortal and immortal. And humans treated them as such: they prayed to them and hunted them. They idolized them and killed them. Animal sacrifice, a rite common to many religions, was an expression of this otherworldly aspect of existential dualism. During sacrifice, the animal is not simply killed or slaughtered; instead, it's handed over in a symbolic act to the god being honored. It's freed of its utilitarian functions and released to a higher order. Viewed from afar, even bullfighting is reminiscent of this type of sacrifice. The human who kills the sacrificial animal always surrenders a part of himself, if not physically then symbolically. In following the rules of a ritualized slaughter, he is submitting to a higher order. A part of himself is always offered to the gods, as well.

Even humans today—who live in an overly technological world, in which real animals are encountered either as domesticated pets or caged specimens at the zoo—can comprehend the reason behind this. Animals resemble humans in many profound ways. The similarities start with anatomy and end with social behaviors. Animals have blood that pours from their wounds. Highly developed species have bodies made of bone, muscle, and skin, like those of humans. Most important, animals have eyes with which they look at humans, and behind which humans can sense an active awareness. In certain circumstances, even the gaze of a carp can be thought-provoking. An animal's gaze can convey both joy and pain. Furthermore, animals can learn. They react to our actions, they adapt to our behavior, they possess social intelligence, and they create differentiated social groupings that are sometimes superior to those of humans.

Still, animals are not humans. They are similar to humans, but also differ from us. This similarity, in the absence of identity, as one might say, is at the heart of the human-animal relationship. Animals are mute. Or rather: they do not speak a language we can understand. We don't know if and how they think, or if they have a conscience that parallels the human conscience. Direct dialogue with them is impossible, even if horse whisperers or dog trainers would have us believe otherwise. In addition, as social beings, humans typically prove superior to animals. When they get organized—for instance as a hunting party—humans can overpower even the wildest and most dangerous of animals using relatively simple means. Prehistoric humans were able to slay mammoths employing the most primitive methods—and these were animals with the obvious advantage over humans in power, size, and speed. This organizational superiority has allowed humans to rule the world, not animals.

At the same time, while severing the tie to these mute creatures through the violent act of hunting, humans never lost this sense of similarity that ultimately reconnected them with the animals. This contradiction, this tension constitutes the curious inner figure of existential dualism between humans and animals. In this sphere, ostensibly contradictory deeds occurred that were, in truth, nothing other than concrete manifestations of existential dualism. Humans could keep animals as companions, yet slaughter them. Animals were hunted, yet shown the utmost respect. This can still be seen today among indigenous populations. The Koyukon are a tribe in Alaska. They are members of the Athabaskan-speaking ethnolinguistic group. The Koyukon live closely connected to nature. When a bear hunter returns from a successful hunt, he waits awhile before finally gesturing toward his take and saying, "I found something in a pit." In doing so, he avoids drawing attention to his hunting success and thereby placing himself above nature. To do such a thing would amount to a breach of protocol punishable by future hunting failure, sickness, or even death. Vestiges of this attitude can still be found in rural areas, where farming has not yet become a fully industrialized enterprise. A farmer can bottle-feed a calf, give it a nickname, and come to love it. The very same farmer can take that calf to the slaughterhouse the following day and enjoy a veal schnitzel that evening. People from the city, who are familiar with the innermost circle of human-animal relationships through movies and books alone—that is to say, who have had it communicated to them through culture— are often dumbfounded by such "heartlessness." They are happy eating meat produced under industrial conditions, and at the same time, they dote on the little calf they meet on their farm vacation. The one thing holding these incongruous feelings together is a sentimental attitude toward nature. "Nature" takes place primarily in the mind—not the nature we have shaped into theme parks to serve the relaxation needs of city dwellers, but nature in its emphatic sense, in which all living creatures are united as an organic whole. Over the course of the last two hundred years, during which time industrialization has dispossessed us of animals, we have preserved nature inside ourselves; we have internalized the animal world, and we think that in doing so, we have actually already saved it. A dangerous fallacy. The farm vacationer who denounces the farmer for having the calf slaughtered has fallen out of step with existential dualism and has slid into a mind-set favoring abstraction, in which animals feature as nothing more than sentimental visual aids.

The vacationer, however, is missing something critical: the unity of humans and animals is not a form of sentimental identification, but something far deeper. Existential dualism is grounded in the human capacity to suffer and to acknowledge the same capacity in animals. The farmer slaughtering the beloved calf suffers vicariously through the animal; in fact, to uphold Judeo-Christian tradition, perhaps the calf is even suffering in the farmer's stead. The identity that exists between humans and animals is not a question of definition; rather, it is a metaphysical mystery based on the logic of pain.

A quick look back at the history of philosophy shows differing opinions on animals' capacity to feel pain. There are two opposing camps: while many authors acknowledge this capacity in animals, others see them as a type of machine, driven by instincts. In the philosophical debate surrounding this topic, a curious tension stands out that is worth examining more closely, because it represents the basis of our current attitude toward animals.

Aristotle was the first to dedicate his time to the study of the animal physique and psyche. He divides the soul into three categories: all living creatures—humans, animals, and plants—possess the *anima vegetativa*; humans and animals are distinguished by an *anima sensitiva*; and humans alone are endowed with an *anima intellectualis*. Augustine describes the animal soul as an "*anima quae non intelligit*"—a soul without intellect. According to the church father, however, animals do possess the powers of sight and memory, which guarantees a sort of active spiritual principle. Augustine calls the animal soul an "*anima vivificans*," which describes a principle existing independently of the body.

In *Critique of Judgment*, Immanuel Kant also highlights a parallel between human and animal actions and draws the conclusion that they are generically related, and that animals must therefore also feel pain. He writes: "What we can quite correctly infer *by analogy*, from the similarity between animal behavior ... and man's behavior ... is that animals too act according to *presentations* ..., and that regardless of the difference in specific kind between them and man, they are still of the same general kind."

It is first and foremost Schopenhauer, though, who accuses humans of not doing animals justice. Animals are our brothers, Schopenhauer writes, but humans in the Western world do not know them anymore; instead, they imagine animals as something fundamentally different from themselves. According to Schopenhauer, the difference between humans and animals is only a matter of intellect, which is to say, of the somatic differences in the brain. European morals did not "provide" well for animals, he argued, and animals deserved our compassion.

Schopenhauer especially targets the rationalistic tradition that began with Descartes and lasted till Jean-Paul Sartre. In this tradition, animals are stripped of any form of conscience, and therefore of the ability to change things in their own world. Descartes says animals have no language, no reason, and no "*esprit*." The "animal soul" is fundamentally different from the human soul. Nature functions in animals like a machine. For this reason, the argument concludes, they can feel no pain. Today, our interaction with animals is still defined by this attempt to prove that we are principally different, thus stripping our compassion for them of its rational basis.

BIOPHILIA VERSUS BIOPHOBIA

The tension between the philosophical views on the topic can be traced back to a fundamental fact: for humans, nature can be a force both nurturing and adversarial. Humans are at once biophiles and biophobes. Edward O. Wilson's central thesis on biophilia posits that humans fundamentally exist in a positive relationship to nature, and instinctively attempt to connect themselves to living systems. This is the reason people put flowers in their homes and go for walks in the woods. Driving this, Wilson argues, is a genetically programmed tendency to want to acknowledge nature as humans' actual habitat.

Close cohabitation with animals also prompted development in the opposite direction; it sparked a process of self-recognition, self-reflection, and of defining boundaries, over the course of which humans came to see themselves as different from animals. Humans lived in unity with their natural surroundings and with animals, and for that very reason, they needed to develop the ability to differentiate themselves from other animal life forms. After all, nature always represented a threat to them, as well.

Humans therefore exist in a negative dialectical relationship to nature. Human reason is of nature, and at the same time, it is not. It is borne of nature, but must separate itself—indeed defend itself—from nature, as a matter of self-preservation. The negative side of the ambivalent state of human reason—being of nature, while being not of nature—was formulated in Adorno and Horkheimer's theory of biophobia in *Dialectic of Enlightenment*. They claim that since the dawn of reason—the beginnings of which were already in evidence in the earliest Hellenic cultural achievements—the relationship between humans and nature has been ill-fated. The idea of humans as rational creatures capable of discourse heralded the start of the exploitation of the animal world. As soon as humans started thinking about their own self-perception, it was all over for animals. They were gradually degraded to objects that humans could control as they saw fit. Even religions scarcely set limits to this exploitation. Animals were first bred, slaughtered, and used for reasons of subsistence—later, for reasons of speculation. The consumer world transformed animals into biomass, into mere suppliers of raw materials. The powers of production unleashed in the Industrial Age perfected this exploitation.

This very interplay of biophilia and biophobia fulfills a fundamental role in the equilibrium between humans and nature. There is one factor, however, that this experience absolutely requires: the actual presence of animals in the real lives of humans. This equilibrium is lost today, because over the course of a two-hundred-year-long process of marginalization, animals have disappeared from human life. All that remains are relics of animals that live in our current world and with which we no longer have a sensory connection—animals as pictures, animals as entertainment, animals as socialized family members.

This process of marginalization even extends to a process of physical deconstruction. Work and production constitute up to 80 percent of everyday human life. In this context, little by little, animals have been replaced by machines. The essential factors responsible for this were inventions in automation that unleashed productivity: the railroad, electricity, steam engines, internal-combustion engines, the assembly line, fertilizers, and so on. While initially used as the occasional stand-in for machines, animals were then increasingly degraded. This radical physical reduction of the animal, as it now manifests itself in the use of animals as test objects, increased steadily in the twentieth century. Biotechnology has managed to disassemble animals into their smallest component parts and then put them back together again. The cloning of animals—that is to say, their gratuitous multiplication—is not only the ultimate form of marginalization. Through cloning, marginalization transforms into the downright deconstruction of animals. The animal changes from the one it was, albeit marginalized, to one of a different kind. The biotechnological disassembly and genetic recreation of animals is the final consequence of that tumultuous moment in the history of human society that could be called the Cartesian threshold. For if, as Descartes taught, body and soul are fully distinct entities, but animals do not have souls, then an existential dualism between humans and animals cannot exist. Animals are then condemned to be machines. And that is exactly what cloned sheep are: living blueprints, extreme forms of the image, but the one thing they are no longer is humans' companions in misfortune.

The French philosopher Elisabeth de Fontenay, who wrote a history of animal philosophy worth reading, points out that reflections on animals, which can be found in the earliest examples of human culture, are nearing their end and will soon be essentially impossible. Soon enough, "animals," as we know them in the Western world, will no longer exist, at least not as the creature called "the Animal," the "Other" that contrasts with humans, something independent and alien. Using the smallest biological building blocks, we can build, change, modify animals. We can manipulate and multiply them. And with that, "the Animal" loses the very last remnants of agency, identity, individuality, and equality with humans that it was still granted in a world that has functionalized and industrialized, subjugated and marginalized its kind. A chicken that endures its bleak existence in a battery cage, pumped full of antibiotics and utterly maladjusted, is still the same chicken it would be if free. It harbors the potential for a new beginning. A cloned animal, meanwhile, belongs entirely to humans, to whom it owes its very existence. A threshold is crossed here that signifies the end of their otherness, their alterity. The genetically "engineered" animal is a pure object, materialized thought, a blueprint brought to life to serve human demands. And thus marks the expiration of the subjectivity and freedom (however they might be described) that make animals the only nonhuman companions humans have. After "the Animal"—if not "the animals"—has vanished from the earth, humans will be left alone, because they will have destroyed the single partner representing otherness with which creation provided them. In giving up this partner, humans lose the possibility to experience the Other and to define themselves in relation to it. They are then reflected back only on themselves, vulnerable to the dangers of total self-reference.

THE FIGHT OVER ANTHROPOLOGICAL DIFFERENCE

In his book *Us and Them: The Science of Identity*, author David Berreby proposes the theory that humans have the enduring tendency to divide their social world into "us" and "them." This does not apply solely to the distinctions drawn between humans, but also to various boundaries between humans and animals. For long periods over the course of human history, animals were the "Other," beings fundamentally different from humans: animals were "them," not "us." Urbanization began with the advent of industrialization in the nineteenth century, and with it humans gained distance from the animal world. Something strange happened in the course of this change. Animals suddenly became "us." They were socialized into human life. To explain this process, historian Richard W. Bulliet divided the history of human-animal relationships into four stages: separation, predomesticity, domesticity, and postdomesticity, the latter describing the phase in which we currently live. What Bulliet discovered: the more closely humans lived with work animals that secured their livelihood, the more likely these animals were seen as "the Others." The further humans moved away from work animals they could exploit for economic gain, the closer their relationships to their house pets grew. Inclusion implies responsibility. This also means that we carry the moral burden that arises from the move to change animals from the Other—from "them"—to one of our own, to "us." A moral dilemma emerges: the more meat we eat (and the less we know of its origins in this modern world), the more shame, regret, or even revulsion we experience, and the more willing we are to welcome animals into our social realm. Vegans with lots of house pets are symptomatic of the era of postdomesticity: witness bestselling authors such as Jonathan Safran Foer, whose *Eating Animals* is a treatise against meat.

Our (Western) relationship with the animal world—which, in addition to primates and elephants, also includes the microcosm of one million known insect species—is stuck in the middle of an animal inclusion dilemma. In light of what we have done and continue to do to animals, we are searching for an opportunity for redemption. We are therefore susceptible to arguments that bring animals closer to us, that allow us to absorb them into our inner circle. In doing so, we like to draw on theories from anthropology and the science of animal behavior, as well as research in brain physiology and molecular biology, that in the last few decades have revealed new, astonishing analogies between human and animal intelligence, between human and animal behavior. The findings seem to confirm the theory of the "us," which we would like to hear. In reading these studies, it can sometimes seem as though there were no anthropological difference between human and animal, between us and them. In truth, the line is razor-thin: humans and chimpanzees share 99 percent of their genetic makeup. Sometimes it can even be difficult to see what difference the remaining one percent makes.

One particular event that gained widespread attention bolsters these occasional doubts. In 1996, a young boy fell into the gorilla enclosure at the Brookfield Zoo in a Chicago suburb and was knocked unconscious. The female gorilla Binti Jua gingerly picked up the three-year-old and placed him carefully in front of the door zookeepers used to enter the enclosure. The scene was caught on home video, and the Internet has ensured its eternal survival. At that time, primate researchers had never seen such behavior outside of the family group. But was this proof that the boundary between humans and animals was an artificial construct?

The academic discussion regarding anthropological difference is undecided, and will remain so. Many thinkers allude to the dangers invited by an assimilating worldview that aims to dissolve the borders between animals and humans. French philosopher Jacques Derrida, who devoted himself to studying human-animal relationships in his later works, considers the people who question these borders to be stupid, plain and simple, because they deny the obvious. Their attitude makes it impossible to recognize the animal as an animal, to let "animals be animal." In this discussion, according to Derrida, "differentiation" is always equated with "hierarchy" and "ranking." Those who want to end the human reign over animals need not deny the differences between humans and animals. On the contrary: only those who accept the experience of fundamental dissimilarity can truly perceive animals as animals.

The unremitting battle over the question of anthropological difference doesn't help the animals on this earth—neither the twenty thousand highly endangered species, nor the millions of animals held on factory farms. Regardless of whether Binti Jua acted more like a human or an ape, and regardless of whether the one percent genetic difference—the one separating us from primates and primates from us—is seen as a distinguishing factor or a *quantité négligeable*, a few things remain the same: the alienation of humans from the animal world in the era of postdomesticity, the "green conscience" that is the sterile product of humans' and animals' drifting apart, and the continual decimation of habitat that does not really affect anyone, because it is happening "elsewhere," and not within the radius of our own reality.

In order for humans to recognize *the very potential for alterity* in animals, there needs to be a connection. This connection is found in the animal gaze. It is in the eyes of animals that humans can see a possible counterpart, a possible Other, separated (only?) by speech and connected by pain. The eyes mirror the pain that binds us to animals. Gilles Deleuze once beautifully described the plane of (co) existence in which humans and animals meet as follows: "This is not an arrangement of man and beast, nor a resemblance; it is a deep identity, a zone of indiscernibility more profound than any sentimental identification: the man who suffers is a beast, the beast that suffers is a man." The gaze of a cloned sheep is different than that of a real sheep. This cloned gaze is empty.

The connection between humans and animals is therefore not a question of choice or culture or a similarity of figures that feel sentimentally drawn to each other and become entangled; instead, it is based on a deeply rooted, kindred capacity for suffering. Is a return to this common ground possible? It is, but only once we see animals as existential companions, and only if this occurs within a context that allows the animal to emerge as an essentially distinct creature, a concrete individual that we can look in the eyes—and find an individual staring back at us.

WHY THE INTERNET IS CRAWLING WITH CATS

THE INTERNET AS A SHARED SPACE OF BEING

It's an unavoidable fact that the Internet is positively teeming with animal pictures. If asked about these little mascots, most people probably think first of cats: cats in washing machines, cats in garbage cans, cats in spacesuits. Cats that look like Anderson Cooper, Tom Selleck, and Adolf Hitler. Flying, swimming, dancing cats. And designer Karl Lagerfeld's darling, Choupette, who has over sixty-five hundred Facebook likes and forty-eight thousand Twitter followers. The Google search term "cat videos" yields some six million results. Why are there so many cats on the Internet? Every Internet user has probably wondered this at least once, without coming up with a plausible answer. Cats seem as ubiquitous on the Internet as pornography.

Cat pictures are members of the genus *memes*, a phenomenon that evolutionary biologist Richard Dawkins introduced to the debate about the evolution of culture as a humanistic counterpart to "genes." Memes constitute the cultural DNA of the Internet. The daily playful exposure to cat memes gave rise to "LOLcat," a language specific to the memes that generate millions of clicks per day, and that resembles something along the lines of an Internet Esperanto. The language could be called "laughing with cats," since "LOL" is an Internet slang acronym that stands for "laughing out loud." Images of cats are overlaid with orthographically incorrect captions and shared throughout the Internet. The LOLcat-encyclopedia *How to Take Over Teh Wurld: A LOLCat Guide 2 Winning* spent weeks on the *New York Times* bestseller list. A LOLcat musical has been written, and even a LOLcat Bible translation is in the works.

While one does not necessarily need to understand it, one needs at least to acknowledge the fact that LOLcat is a pop cultural phenomenon of the highest order. But is it worth spending any time thinking about? Or are these memes nothing more than evidence for how quickly cultural degeneration and depletion progress when technology runs on automatic, independently of humans? The speedy dissemination of memes is only possible because there are websites that allow these images to be created quickly and easily. These sites are visual vending machines. Memes are little more than industrial artifacts, cheap icons of the Digital Age that are produced without thought.

Of course, this very "production without thought" is not insignificant. It stands for something, signifies a state of affairs. Therefore, there are logical and earnest theories about memes and the Internet's "felinophilia" that are interesting to consider in connection with the Animal Internet and what it can constitute. For the theories suggest that cat pictures on the Internet are indicative of the search for existential closeness to living creatures. One theory claims that looking at cat pictures can reduce stress. Another hypothesis states that cat owners have fewer social contacts than dog owners, because cats don't need to be walked. In order to share in the enjoyment of having animal-centered social contacts, though, cat lovers are quick to share pictures of their animals on their social networks, which allows them to become friends with other cat lovers. Online chatting replaces sidewalk encounters. A third explanation draws on the symbol of the cat in human cultural history. Cat images can be found in many cultures. They're often a symbol of freedom and independence. Humans share cat pictures online to express their inner desire for freedom. These cat images could then be seen as small, everyday acts of rebellion and declarations of independence in an online world increasingly defined by surveillance and control. Finally, another theory provides a psychological explanation for digital animal pictures in general. This theory proposes that humans using the Internet—that abstract structure, that cold space—miss the living world. Humans appear to have the propensity to want to make the Net more human, to fill it with organic content. The animals of the Internet thus function as a means of "softening" the Net.

FROM SPACE OF BEING TO LIVING SPACE AND BACK AGAIN

As we said: the significance of cat pictures on the Net shouldn't be overestimated. They may, however, be a sign of our shaping the Net into a space for speech and contact, thus serving as a precursor to the shared "space of being" taking shape for humans and animals on the Internet. Images of nature today are usually digital and accessed virtually. They're no longer experienced; instead, they're downloaded, consumed as something prepackaged and one-dimensional. They no longer penetrate the textures of our culture. The proof that this was once different can be found in humanity's biophilic memory. Examples include the prehistoric cave paintings in Lascaux, ancient myths, European fairy tales about the rabbit and the hedgehog, La Fontaine's fables, and countless other works of literature and art, but also everyday expressions like "fight like cats and dogs," or "smell a rat." Fairy tales reflect this symbiosis. Humans and animals encounter each other on equal footing. They communicate with each other using the same language. This language is both artistic expression and the reflection of a shared space of being. At the time these fairy tales emerged, animals were both humans' closest companions and their most dangerous adversaries. They defined human existence, both the good and the bad. It was only logical that they should be assigned central roles in these texts and tales.

Animals today are no longer as present in literary texts. This is symptomatic of a fundamental transition: talk today is only about animal living spaces, and no longer about a shared space of being between humans and animals. This shift in language is telling. The difference between a living space and a space of being is that, in the former, animals and humans inhabit mostly separate realms; we each have "our place." In a shared space of being, however, humans and animals are dependent on one another and commonly live in coexistence or conflict. However, as opposed to separate living spaces, there is constant contact. The language the wolf uses to speak in fairy tales is the symbolic expression of this ontological closeness. In comparison, the wolf that strays from the zoo and onto city streets today comes across as a foreign body in the metropolitan living space. It has escaped the space humans had assigned it.

As discussed earlier, the image of nature, the "idea of nature," has developed in tandem with the disintegration of the space of being into living spaces, or as they are better known today, habitats. During the Industrial Revolution, the idea of nature served a compensatory purpose as a collective memory. Nature increasingly became the antithesis of the mechanical and utilitarian world: untouched, fertile, free, undemanding, pure, and ideal. Nature was no longer simply "there"; rather, a new sense of protective duty arose to beget nature "as such." The genesis of the idea of nature is also the birth of ecological thought.

The first area that illustrated this paradigm shift from nature to the "idea of nature" was gardening. While French gardening theory could easily posit, well into the late eighteenth century, that the ideal was achieved through the architectural arrangement of the garden, following the French Revolution, the theory of the English garden broke from the dogma of the man-made and called instead for nature's liberation in the form of more open, authentic designs. Symmetrical perspectives were ousted by asymmetry and surprising moments that were meant to mirror natural phenomena. Naturally, the English garden was also the product of precise calculation. It was even more artificial than the French garden, because it did not admit to its artifice, instead simulating nature. The French garden uses natural elements to create a self-aware artwork, while the English garden attempts to conceal the character of the artwork and create the impression that it has reconstructed nature. The idea of nature, as formulated and refined in the English garden, is also the first step toward the abstraction of the natural. Von Humboldt and Goethe took this step. They were both gifted observers, but their observations were based within an aesthetic system that related more to themselves than to nature.

The "idea of nature" was a product of the nineteenth century—an understandable and necessary product. The twentieth century, however, then produced the "*idea* of the idea of nature." Slowly but surely, this came to replace the "idea of nature." The "idea of the idea of nature" is the memory of a memory. It has distanced itself even further from the matter; it is an abstraction in the second degree. The "idea of the idea of nature" concerns itself with derivatives, and no longer with the matter itself, as was the case with the "idea of nature." Goethe still pasted dried plants in his diary and made animal sketches, in order to understand nature and get closer to its phenomena. Today, all we want to know is how nature "works." A modern-day Goethe would ponder population numbers and migration dynamics, and he might make sketches of biochemical connections. He'd have little reason left to concern himself with nature's concrete elements. And even Goethe would have trouble poeticizing this second tier concept in a way that would move us; the distance to actual nature identified by our biophilic consciousness is simply too great.

The "idea of the idea" governs our behavior toward plants, animals, and landscapes. This expresses itself in actions that no longer emerge from a concrete interaction with nature, but from a derivative, symbolic value system. We pay attention to nature only when it's threatened or when we want to make a social statement, perhaps because of a lingering guilty conscience, but definitely because of distance. We're still aware of the responsibility we have for creation and its diversity of species, because the idea of the idea of nature is constantly evoked in politics. Because we have fallen out of the shared space of being with nature and animals, however, we pass the responsibility on to experts. This abandonment of primary responsibility is emblematic of the era of "the idea of the idea." Nature—the most universal and accessible aspect of creation—has become a specialized discipline. This is the central tragedy of the "idea of the idea," which also applies to other social topics. In politics today, key players allow themselves to be guided by "ideas of ideas" that have largely lost their concrete reference point in reality. We get involved with nature in the same way we vote in elections—spurred by tedious mobilization campaigns. We may make the odd donation, thereby participating symbolically in conserving nature and saving animals. An inner connection to the cause we're supporting, however, or a clear idea of what happens with our money and what exactly we're advocating, has more or less ceased to exist. Even today's doctored nature images barely manage to touch us however spectacular the pictures may be. Not even the bloodiest Japanese whaling scenes can move us. Once the evening news has ended, *CSI* has begun, or we lose ourselves surfing the Net.

THE NEW LANGUAGE FROM HUMANS TO ANIMALS

The only thing that can help us escape the prison of "the idea of the idea" is to develop a new language that overcomes distance and creates closeness. Above all, in the new relationship with animals, it is critical to build emotional connections. The key now is eye contact, as opposed to mere observation. The concept of closeness has shaped the structure of the Animal Internet. In the past, animal behavior could be observed only by those in close physical proximity to wildlife. Today, technology allows for the global monitoring of animals: global proximity. For the time being, however, this closeness is based on data and is not yet linguistically structured. It comes across as random and abstract. Knowing where an animal is at any given moment is not yet the same as feeling close to that creature. In order to start growing closer to animals, beyond what the data enables, the personalization of animals is necessary. The animal data must come to form a life story—a destiny, even—that has highs and lows, light and dark moments. The animal becomes approachable once it, too, is seen as having a fate. Only then can the shared space of being become a space of shared language.

This linguistic space is an ancient human dream come true, namely the wish to share a common language with animals. We can more or less get simple points across to them, especially when they are our house pets. We cannot actually speak with them. With the help of sign systems, we are able to cultivate a sense of confraternity with some of the more advanced species, such as dogs, and we can tell from their behavior that they are responding to our signs, whether obeying or defying us. This is, however, a far cry from anything approaching the verbal exchange of the content of consciousness. It could better be described as a form of conditioning that is complicated and difficult to learn, and that requires continual testing and repetition of the simplest of signals. Furthermore, this communication is essentially one-dimensional. It tends to move from humans to animals, and not the other way around.

Similarly, we are also able to observe the sign systems animals use to communicate with one another, and to speculate about the "languages" that seem to be represented there, from whale songs to bees' "waggle dance." We cannot, however, speak this animal language, which can do remarkable things, and about which we don't know enough to determine how close it may be in quality to our own language system. The signs animals use to communicate with one another have nothing in common with those we use in attempts to interact with them. Apparently, creation has disintegrated not only into two spaces of being, but into two language zones that also remain unconnected.

We certainly see this inability for a true exchange with animals as a shortcoming. We can only speculate as to whether animals feel the same. It's not impossible that animals want to say more to us than they can. One could certainly get this impression from observing the behavior of dogs when they approach humans for a purpose, but without linguistic tools to express their exact goals: Do they want food? Closeness? Our attention? Might they even be trying to warn us of something? While dog owners don't often understand what their dog is trying to tell them, they are, however, often convinced that they know their dog and can read its gestures. This inability puts us in mind of the split in creation that has left humans detached from their origins. Indeed: in the Bible, humans and animals were joined by a common tongue in Paradise. How else could the serpent have spoken to Adam and Eve and tempted them? The loss of a shared language also symbolizes the loss of the state of nature.

Thus the dream common to all human cultures, to be able to understand and speak the language of animals. This dream is another form of biophilia, while the desire to control house pets through commands displays clear biophobic characteristics. Stories about humans who can speak the language of animals arise throughout history, from King Solomon to Dr. Doolittle to Buck Brannaman, the inspiration for fictionalized horse whisperer Tom Booker. In ancient myths, this ability stands for godlike wisdom and sagacity, and today it has the aura of esoteric lore. In all cultural constellations, however, it's a sign of human recognition of the Other in animals, seeing them as essentially different creatures against which we orient our own existence, and to which we seek sensory, individual access. This wish, however, stands in opposition to the species-centric thinking that has largely shaped our learning about nature, and that dismisses any form of individualization of wild animals as "humanization." Humanization is an anthropomorphic construct and therefore represents disjointedness in the human-animal relationship. For decades, "habitat-centric thought," which developed logically from "species-centric thought," was seen as a triumph over the "unscientific" *Flipper* ideology, which was considered inferior because of its humanizing approach. Backing this was the conviction that by abandoning this anthropocentric perspective, animals would gain freedom and autonomy with regard to humans. The very opposite has proven true. Animals have fallen further into the margins, because people have lost sight of them.

What does a swallow that has been swept into a storm

experience, or better yet, feel? If and when we are able to answer this question, then we will know we are dealing with an animal Net personality, with whom communication is possible. How exactly does this kind of wild animal Internet personality—complete with its own fate and with whom we can speak—emerge? To start, there are the recorded raw data that the tagged animals provide, which may also include physiological readings. Data like heart rate and blood pressure can provide conclusive information on the animals' behavior in any given situation. In gaining insight into their cognitive inner life, we must consider the impact on animals' subjectivity, while at the same time grappling with subjectivity itself; indeed, we are required to concede that animals possess the capacity for subjective experience. By aligning animals' internal status with the external situation they are currently facing, we are granted insight into animal decision-making processes in the wild. "Patterns in animal movements allow us to understand how they think and how they reach decisions in concrete situations," Martin Wikelski notes. A series of tangible moments gradually accumulates, until a complete animal personality and character form. Every animal individual will make decisions and behave differently in comparable situations. When taken all together, a life story emerges from these decisions, even if the animal has nothing more than a number for a name. For example, we know that Stork Chick 2539 was born on June 7, 2012; that it chose to nest in Dannefeld, west of Berlin; that it was tagged at the ornithological institute on Lake Constance in the south of Germany; and that after getting caught in a sandstorm in North Africa in the autumn of 2012, it landed, near blind, in the Moroccan oasis city of Errachidia, where a young girl cared for it devotedly, but where it nonetheless died on November 14, 2012.

This personalization brings about wholly new situations in human-animal relationships, especially with so-called "nuisance" wildlife. This was the case with one wolf in Yellowstone National Park that did, at any rate, manage to get an obituary in *The New York Times*. In December 2012, after Wolf 832F was shot outside Yellowstone—that very year, wolf hunting had been sanctioned again in Wyoming—a deluge of outraged phone calls, emotional postings, and wild online commentaries ensued. As it turned out, Wolf 832F— the alpha female of the Lamar Canyon wolf pack—was outfitted with a $4,000 GPS collar that monitored the animal's every move. Her life could be followed in fairly close detail on the Internet. The American public knew not only her, but also her kin, her pack, her pups, her hunting patterns. In short: the life of Wolf 832F was a public event. She was soon the most famous wolf on Earth. The pack led by this shewolf very rarely crossed park borders, as evidenced by the GPS data. This one time, however, the alpha animal wandered over the line, and immediately, things went awry. *The New York Times* used this as an occasion to print an op-ed piece on the wolves of Yellowstone. And thus, Wolf 832F became a VIA—Very Important Animal.

Yet another American wolf recently launched its online career. On June 8, 2014, *The Washington Post* printed the headline, "A Lonely Wolf Gets a New Mate, Powerful Friends, and a Little Protection." In this case, too, a hitherto unknown wild animal received prominent coverage in a major newspaper. At the center of this story is

OR-7, born in 2009 in northeastern Oregon and collared in 2011. OR-7 abandoned his pack in September 2011, and in the following years, he traveled over two hundred miles from Oregon to northern California, in search of his own territory and a mate. They called him "the wandering wolf." The case is remarkable in many ways. Wolves typically hunt in packs. A wolf that has lived alone for years must be strong and experienced enough to hunt all its own food. In his wake, OR-7 left chewed-up bones of the elk he had single-handedly killed. In addition, remote camera photographs documented that the wolf was, indeed, successful in finding a mate: a female with a black coat. The wolves never appeared together on camera, but they were caught in close succession, which suggested that they were a couple. This wolf's life, which is not yet over, was retold in a short film that elaborates on his personality. This was a necessary course of action, because the return of wolves to California is still relatively fresh. Before their recent return, the last known wolf was killed in 1924. Farmers in northern California are less than thrilled by OR-7's appearance. The California Cattlemen's Association has spoken out against the state wolf management plan, which aims to protect the animals. This battle has not yet reached its conclusion. It's uncertain whether the conflict will be resolved in court or by politicians, because California's forests are vast, and weapons are not exactly scarce in those parts. The fact that this animal is a known quantity, whose life is monitored closely via video and GPS data streams, could be an effective deterrent.

We need something more than subjective data to understand the relationship between the New York businesswoman dragging her groomed and costumed little Chihuahua on a retractable leash through the street canyons of Manhattan, and lice-ridden, filthy OR-7. The blood sugar levels of a Galápagos tortoise aren't exactly known for sparking great conversation, either. Data alone are not enough to dismantle prejudices. That can only be accomplished by sharing experiences and emotions. Both the wolf and the turtle need stories in which humans can take part. Animals will return to the human realm by telling their own stories, writing their autobiographies. The biography of Sicklefin, a young shark (male, 1,724 pounds, just over twelve feet long) whose escapades are captured in real time on the flat screen is no fabricated tale; it's a glimpse of reality that can be followed with the Shark Net app.

To experience the animal world with heightened sensitivity, powerful images and sounds are needed, in addition to the story lines. If animals' surroundings are meant to become truly vivid, then humans must be able to dive into the dark reaches of the ocean with humpback whales, fall upon prey at breathtaking velocity with a falcon, or fly majestically over the peaks of the Alps with an eagle. Researchers have already generated these images by equipping the animals with tiny video cameras and uploading the spectacular images onto YouTube. The result is an entirely new kind of sensory experience that allows us to see the world from the animals' perspective.

Many life stories of actual animals already exist online, such as stork diaries kept by Germany's Nature and Biodiversity Conservation Union or *National Geographic*'s GeoStories, which use data gathered by Movebank.org— the Max Planck Institute's online database of animal tracking—to tell the stories of sea turtles and bald eagles. These digital animal tales are still in their infancy, but wherever this method has already been systematically introduced, the relationships and attachments that develop between the humans and animals are remarkable. Even the stork diaries inspire a lot of comments. The waldrapp Facebook page, however, remains one of the best examples of this, especially because the profile is enriched by photographs and observations shared by the animals' friends, who have encountered them in the wild, observed their behavior, and then tell their story.

Projects like these also make it clear that animal stories need not be told by a single author or editor, who simply draws conclusions from the available data, and whose stories may toe the line of fiction. The Web offers the opportunity for crowdsourcing. These animal stories can have many different writers, as people who have spotted the animals and shared their observations become coauthors. This collective storytelling has three positive aspects: first, it concretizes the story being told; second, it motivates people to get closer to animals and become invested in them; and third, it promotes human togetherness. Future technologies will even allow animals to blog autonomously, that is, without human help. Computer systems can already supplement an animal's positional data with other information about its concrete location, such as details on the habitat type and local weather. Furthermore, the animal's conduct there can be compared to its territorial behavior, thus determining if the animal is at home or not. These data are then fed into a text generator that creates a blog entry telling the tale of a specific day in the life of, say, a red kite (*Milvus milvus*).

These eloquent red kites are not labeled with numbers; instead, they have individual names. Their names are Wyvis, Moray, Millie, and Ussie. Scientists are still in the habit of avoiding personal animal names, in order not to unnecessarily emotionalize their work. They tend to use mixed abbreviations of letters and numbers. But that's half-hearted. Names are no lark. It's in naming a wild animal that we come to view its life as a destiny in which we take part. It's at that moment that the animal becomes identifiable as a unique creature. This was particularly the case for Tolosa, a five-year-old, female brown bear living in the forests of Slovenia. We know more about Tolosa's life than about any other brown bear on Earth. For one whole month—in October 2013—Tolosa wore a collar equipped with a GPS transmitter and a video camera. The devices recorded the animal's daily activities from a first-person perspective. For a month, five minutes of footage were recorded every hour. Researchers at the University of Toulouse purposely chose a female for this experiment, since male brown bears are generally much more aggressive. Slovenian researchers had previously only seen Tolosa, who lives in a nature preserve, indirectly. Through the recordings, they were able to determine that she sleeps for much of the day and becomes active at night. Brown bears are typically diurnal. In this region, however, where humans also live, they have altered their own rhythm in order to avoid contact. The footage also revealed behaviors scientists had never seen before. Tolosa was repeatedly seen uprooting trees for no apparent purpose. The zoologists' interpretations were divided. One thesis suggested that the bear was engaged in play, that she was in training, even. The recordings are projected onto a large screen at the Muséum d'histoire naturelle de Toulouse, and are now available on YouTube. They serve to represent the reality of the bear's life, thereby demythologizing the animal. The hope is that French visitors will develop feelings of tolerance toward the bear, because even in the French Pyrenees, conflicts have been known to arise occasionally between predatory animals and humans.

Ecology thus profits from humanization. Josef Reichholf predicts that the Animal Internet will mean improved concrete engagement in nature conservation. His theory states that only those who are emotionally invested will act: "There is no path leading 'back to nature.' What most certainly does exist, however, is an increased orientation toward and attention to nature. Appeals are practically worthless. Experience has taught us that. What really counts is the experience. How the experience is had doesn't even matter that much. If the Internet, with its virtually limitless possibilities, were also to enable more contact with nature, then I would have to welcome it. It's certainly better than our current approach to conservation, which does all it can to keep humans at a far remove from nature, categorizing them in broad strokes as destructive and burdensome. People who will at least engage emotionally with animals on a digital level, are more likely to be willing to do the same in real life, as opposed to those who are barred from the natural world by conservationist measures."

These animals are no longer abstract numbers and faceless representatives of their species; instead, they are creatures of flesh and blood, whose presence demands our compassion. An animal has the power to affect us deeply, because through the concrete example of one individual, we are suddenly able to understand the problems facing its species. A single animal individual can thus become an ambassador and elicit new protection measures, as was the case with Wolf 832F. After the animal was shot in Yellowstone, the National Wolfwatcher Coalition launched a successful fund-raising campaign to raise money for wolf research.

A similar case occurred in Germany where tagged wolves are opening doors for ecological thinking. Nature and Biodiversity Conservation Union wolf commissioner Bathen is already seeing positive results: "All the possibilities online have led to increased general education in ecology. Ecological principles can be better understood when you're familiar with the real life of individual animals." Bathen's suspicion appears to have proven true. When a wolf was shot illegally in the western Westerwald region last year, pictures of the dead animal spread across Facebook like wildfire. Sharing became caring, functioning here as active nature conservation. The media picked up the images, and the widespread outrage resulted in the hunter being stripped of his hunting license.

MARTIN BUBER AND MAYA THE BEE

After thirty years of nature and wildlife conservation, access restrictions, and red lists of threatened species, the suggestion that we start addressing brown bears as acquaintances can make us cringe. Only yesterday, we hardly dared to pick the flowers growing alongside the hiking trail, because they might be protected species, and today we're expected to make friends with red kites and elephant seals?

With the advent of animal social media, the literary

genre of animal biography, which arose in timeless works like *Maya the Bee*, is moving into a new phase. Facebook has changed people's relationships with one another. It is now in the process of changing human-animal relationships, too. This is worth closer consideration in an age when the new Pope selects the name "Francis"—Saint Francis being the patron saint of animals and ecology. The Franciscan framework of empathy includes compassion, or *compassio*, for the earth's creatures. Many historical accounts of Francis of Assisi describe him rescuing animals or lovingly approaching them. In *The First Life of Saint Francis*, Thomas of Celano depicts the way in which Francis cared for seemingly insignificant creatures: "Even towards little worms he glowed with exceeding love .... Wherefore he used to pick them up in the way and put them in a safe place, that they might not be crushed by the feet of passersby."

This *compassio* informs the living relationship between humans and animals. The difference between an "I-It" and an "I-Thou" relationship, as introduced in Martin Buber's *I and Thou*, can help elucidate the transformation of an animal from an object of observation to the subject of a relationship: "As experience, the world belongs to the primary word I-It. The primary word I-Thou establishes the world of relation." The animal world, which we perceive as a structured collection of species, belongs to the world of experience. They are—to reference Buber once more—an "object" in our experience of the world, and not a "presence" in a relationship. A counterargument immediately presents itself, however: We can find evidence of the Buberian Thou in our house cat—but in a bird living in the wild? To what extent can such an instance that changes the animal from an It to a Thou be called a relationship?

Social media—with its power to personalize and emotionalize, as well as with the sensory range and closeness it enables—can certainly build hype around the idea of viewing an animal as a true Thou, freeing it from its role as a mere object of observation. Through the images it contains, the Internet generates empathy, which is the foundation of social relationships. Josef Reichholf, too, believes in the Internet as a visual force that creates relationships between humans and nature. He sees an emotional progression from static photographs to the real-time moving images that wildlife cams bring to our computers or smartphones: "A polar bear leaping onto a moving ice floe is, without a doubt, going to be more effective than any and all of the climate researchers' appeals and diagrams that no one understands. Conservation efforts have long used animal images to drum up sympathy. But to be on the edge of your seat, watching the live stream to see if the animal survives, gives rise to a new order of emotion, much more so than could occur through even the most powerful words or photos."

The Animal Internet is the medium of a new nearness, which humans must first examine, and to which they must then adapt. It is assuredly still in the beginning stages of a more or less lengthy phase of enculturation, during which we must decide how close animals are allowed to come to us, how close we are allowed to get to them, and how willing we are to take on the responsibility inherent in this new dynamic. This much can be said about the projects now under way: once contact has been established with one of these individual wild animals, a sense of responsibility for its well-being is quick to develop, and we feel prepared to take a stand for it—and thus for its species and its living conditions. It's a short distance from being a waldrapp's Facebook friend to becoming an active supporter of its reintroduction program.

TRANSHUMAN LANGUAGE?

This form of nearing and nearness obviously does not represent real "speech" with animals. It's not actual dialogue. The language barrier remains intact. Communication between humans and animals is made possible only by the machines or programs that use artificial intelligence from data to make connections and tell stories. The Internet is a fabulous interface, but nothing more than an interface, and as such, a façade that interrupts. Can it be more? Can it open the door to a language that connects all living creatures, a language that could even be called "transhuman"?

The natural philosopher Henry David Thoreau intuited the discovery of a transhuman language in technology, albeit metaphorically, but with such intensity of foresight—indeed, with a real sense of living in the future—that his words are imbued with great truths. On September 22, 1851, on one of his countless perambulations through the woods, he comes across one of the newly erected telegraph poles that mark the introduction of the era of modern communication in the wilderness of New England. He presses his ear to the pole and hears a distant music—the hum of the distant cities that the telegraph connects. It is a revelation for Thoreau: nature and technology coalescing into a new unit. He therefore invokes a tenth Muse within his own pathos, the "Muse of Communication," as he names her, and he envisions a cable encompassing the world and creating a new community for all living things:

"And that the invention thus divinely honored and distinguished ...—is this magic medium of communication for mankind! To read that the ancients stretched a wire round the earth, attaching it to the trees of the forest, by which they sent messages by one named Electricity, father of Lightning and Magnetism, swifter far than Mercury, the stern commands of war and news of peace, and that the winds caused this wire to vibrate so that it emitted a harp-like and æolian music in all the lands through which it passed, as if to express the satisfaction of the gods in this invention. Yet this is fact, and we have yet attributed the invention to no god."

Myth and utopia intertwine in a cyclical figure. Everything humans produce leads them back to something they have lost, abandoned, or destroyed at a different point along the way. The history of nature is not one of spiraling and irrevocable destruction; indeed, there is reason for hope. Nature always finds new ways to speak to us, and it always creates new methods of communication. Thoreau's journal is a moving work, as it conveys an intense dialogue with nature, listening in on the world and writing down what is heard. His relationship with animals is also shaped by the desire to communicate—both to share and to receive. His immersion in nature has religious undertones: communication becomes Communion and participation in a holy space.

Nevertheless, Thoreau can sense the disjuncture in this space. After all, his language is insufficient in describing anything beyond that which bars him from attaining oneness with the natural world. His language is his alone, and the attempt to describe the latter—the force separating him from nature—therefore throws him back to his own individuality in a radical way. Language is also an interface. The attempt to depersonalize oneself through writing ends in a eulogistic evocation of one's individuality, of one's subjectivity. Language is culture, it is artifact, and even if Thoreau did choose the open form of the journal, which does not bind language to rigid structures—instead, giving it the freedom to follow subtly the contours of phenomena, to breathe in time with episodes encountered, and to allow space for sketches—even this language creates a wall that hermetically separates the author from the natural world with which he yearns to unite. And because he experiences and suffers from this dialectic of biophilia and biophobia, of convergence and self-assertion, he sees the tenth Muse— the Muse of technological communication—as a way out of the dilemma. She represents a rebirth. Through her, Thoreau presages a global closeness that encompasses all living organisms. In short, he foresees the liberation of humans trapped in isolation, through communication technology. Thoreau foresees the Internet.

There are people today who still think along the utopian lines Thoreau originally drew, and who believe in the truly utopian communicative potential of digital technologies. In doing so, they naturally depart from the mystical flights of the nature writer. They want to know if it might someday be possible to use a brain-computer interface to translate the content of an animal's consciousness into human language. A year ago, musician Peter Gabriel, of Genesis fame, and Vint Cerf, known as one of the "fathers of the Internet," announced the founding of the Interspecies Internet (I2I), a think tank with support from the Massachusetts Institute of Technology, among others. The I2I is slated to become an Internet that spans all species and enables communication between humans, animals, and other intelligent creatures, potentially including those living on other planets: "We're beginning to explore what it means to communicate with something that isn't just another person ... All kinds of possible sentient beings may be interconnected through this system," Vint Cerf says of the organization's goals in the TED Talk linked on its homepage. In this case, the Internet serves as an interface between different *forms* of consciousness. Peter Gabriel has already used I2I technology to play music with bonobos. Psychologist Diana Reiss, another cofounder of the platform, uses mirror experiments to test whether dolphins or elephants possess self-awareness, which would be a precondition for attributing to them the human faculty of speech. The next step would be to employ a software interface to translate the language of dolphins into a human language—and the other way around.

These initiatives are a logical continuation of the Animal Internet. They operate under the assumption that it could very well be possible not only to enable humans and animals to communicate with each other, but even to develop a bidirectional language between humans and nonhumans. Required here would be the liberation of animal language, the creation of a means of articulation, and translation. This liberation of animal language would be the next step after the liberation of their characters, which we can already see happening today.

WHY AFTER NATURE, NATURE WILL STILL EXIST

HUMANS AND ANIMALS IN THE ANTHROPOCENE

Everything reduplicates. After the first life came the second life, and after the first screen, the second, which means viewers can now tweet live comments on the TV shows they're watching. And now after the first nature, a second nature is due to emerge. What kind of world are we headed toward? Can something like a second nature even exist, which simultaneously mirrors the first and engages it in dialogue, thereby coming full-circle, so to speak? And what will this kind of "nature after nature"—the contours of which are on the horizon, and the first concrete phenomena of which we are already experiencing—look like?

The Animal Internet represents a central element of a planetary digital culture that is fundamentally altering our reality—both "real" reality and the one we experience. It is as much a part of the "digital turn" as big data and Spotify, the NSA and the privacy debate. We must accept the task of developing appropriate Net-specific ethics, along with new morals of nature. This discussion is not making any headway, though. It has gotten stuck in a one-sided argument that represents the polar opposite of the euphoria with which society once welcomed the dawn of the Net. Many people are withdrawing from the Net because they believe it is the only way to preserve their real life. They reject social networks and contrast them with the image of authentic friendships and relationships. Why do they do this? Some simply have not found a way to integrate the possibilities of digital life into their world, because they think they must decide between two competing versions of life. This is also because the industry insists on continual acceleration and innovation, rather than offering technologies that deepen the intensity of the online experience and promote, or even permit, a culture of digital dwelling.

For this reason, because the world is suffering from digital overheating, the discourse about the digital transformation and its consequences has reached a standstill. It has not taken mental hold where it ought to be occurring, which is above all these things. The debate surrounding the Animal Internet can help us adopt this standpoint, or at least aim for it. It can help us see the digital revolution and the media world in a new light. It can deliver the impetus for us to situate our beings in the digital realm.

It helps to look back at a similar discursive threshold that was crossed in the not-too distant past: the moment postmodernism arose from the ruins of modernism. The postmodern created the distance from modernity that was needed to identify modernity as "modernity," in the first place. Humans first freed themselves of modernity—that is, of the namelessness of a relentlessly forward-moving and personally destructive fate—when it became clear to them that they were already living in an "after" that they dubbed "postmodernism;" that is to say, when they realized that the factors that had changed modernism now constituted the very framework of their lives. Strictly speaking, this move was little more than a change, a shift in perspective. It was the liberation of sight. This "little more" is, of course, a great deal more: an altered point of view. The reorientation of perspective is one of the most difficult human undertakings, because it requires humans to rid themselves of conditionality and to defy the reflexes and instincts that tell them to see the new, unknown, and unsettling as potential dangers and threats. In evolutionary self-defense, the human brain has developed to anticipate disaster, to be prepared for the very worst eventualities. The species would otherwise never have survived up to 2016. Within civilization, this expectation of disaster is often superfluous and counterproductive, as in cases when it is not directed toward concrete threats. Viewed in this light, of course, postmodernism was a genius, anticatastrophic human "invention of a concept"—a coup of sorts.

Similar forces are at play with the digital and the postdigital. Liberation amounts to developing a postdigital viewpoint. It must become clear to us that we are no longer living in the digital era, but in the postdigital era, in which our lives are so strongly shaped by digital technologies, that we can no longer remove them from our lives, because they have, in fact, become our lives. This insight results in an enhanced view: in the postmodern age, the pressure for change that had defined modernity was no longer viewed as an inevitable burden, but as a game and a lifting of the rules; similarly, from the postdigital perspective, instead of the digital turn representing a threat, it appears as an expansion of human possibilities. In the postdigital world, the digital is part of our life, of our daily duties, of our bodies, even. It no longer represents something foreign to us. The digital has been overcome, primarily through its availability. The actual digital revolution with which we should be concerning ourselves is not about the digitization of our world; instead, it is about what digitization is making of our lives—how it frees and mobilizes us, how it allows us to experience time and space, and how it is redefining social participation and public domain and much more.

The discourse we are currently having regarding big data, personal freedom, and privacy is not about the effects—it's about the causes. It is a discourse based on the understanding that we are still living in the age of the digital revolution. It focuses on feasibility, on technical facts, and it is defensively formatted and reactionary; it seeks safe havens as is typical in revolutionary times when foundations of security are put in jeopardy and the belief is perpetuated that we must protect what has already been lost. What it needs to be, instead, is offensive, aggressively asocial, innovative, and inventive. Most important, a postdigital dialogue must give up on the idea of how to control the digital, just as the postmodern dialogue did not attempt to domesticate and define modernism retroactively. The essence of postmodernism was not an attempt to distill modernism; instead, it aimed to play with combinations of the elements it had released. The postdigital approach is similar: big data will never be domesticated; rather, postdigital dialogue will relinquish its demand for domestication in favor of a flexible, resourceful, and resilient standpoint that will then lead to a new human freedom. In order to overcome the challenges of modernity, modernity had to be conquered. In order to overcome the challenges of the digital, the digital has to be conquered.

The Animal Internet is a logical extension of this school of thought "beyond the digital." In order to save nature, we must leave it behind. The postdigital era also features a corresponding new geological age. We no longer live in the Holocene; instead, we now find ourselves in a new geological age known as the Anthropocene. The Anthropocene is the "age of the human." In the Anthropocene, humans are the force shaping nature, and no longer the other way around. The power dynamic has gone into reverse. In the last two hundred years, humans have changed the earth so drastically that it is truly valid to identify it as a new geological age. Farming, urban development, and road construction have radically transformed the earth. The Anthropocene is an epoch in world history in which human activity defines all the planet's ecological and geochemical processes. This control will continue to improve human prosperity on Earth. It does, however, have a monstrous dark side: it has a destructive power capable of eradicating organic structures that have developed over millions of years. The earth's destiny definitely lies in human hands. This alone supports the fact that humans cannot leave nature to its own devices and simply hope it will develop in the right direction, provided it's given enough distance from human activity. All indicators show that the current method of granting nature sovereign spaces is only a temporary—and therefore illusory—way of preventing nature's destruction. It's an irreversible fact that these free spaces will continue to shrink, the more human prosperity, and the resulting global demand for it, grow.

NATURE AFTER NATURE

The Animal Internet is thus a symptom of the Anthropocene. The age is characterized by a deliberate pervasiveness of technical structures on the one hand and natural realities on the other, as well as by the influence humans exercise over nature. The Anthropocene pictures nature as a system that has been lodged in human activity. It pictures it as an embedded system. An embedded system of this sort is comprised of both hard- and software components. Both run seamlessly into the other. In the Anthropocene, humans are the hardware and nature, the software. In order to function, they must be aligned, and this is possible only when interfaces have enabled communication—that is, when engineers have programmed the components to work together. The natural phenomena that emerge are the result of a blueprint, a design. The development of nature becomes planned, not in a way to better enable its exploitation, but in order to find the ideal constellation between the two systems. This systemic notion sounds threatening and alien, but ultimately, the integration it effects is reminiscent of creation myths featuring humans and nature in a primeval, unified form. Nearly all of these myths focus on the unity of humans and nature. At the same time, the idea of embedded systems robs nature of the very quality that has defined it over centuries of human civilization processes: the moment of contrast, the characteristic of otherness. Nature was always the Other, ecology defined itself as the attempt to organize our coexistence with this Other, and culture amounted to what could be gleaned from the Other and presented as one's own. This dialectical process comes to a halt in the Anthropocene. Nature can be returned to itself, it can grow into human society; ecology is no longer a discipline tasked with formulating rules of separation, but one that creates closeness, and a new form of images emerges from this closeness, appearing in the context of a grounded culture that has resulted from an elementary reunion with nature and animals.

In the Anthropocene, we encounter nature after nature. What exactly does this mean? In the age of nature after nature, we move through the natural world as through a park. Machines show us where to find which animals. Software tells us which animals we're approaching, which are approaching us, and which are already there, but not yet visible. These new tools of nature even tell us which animals not to expect. Nature is becoming predictable. To be sure: the element of surprise that is so closely tied to a romantic experience of nature—indeed, what defines this experience—is lost. When snorkeling in the Mediterranean, the overwhelming wonder one feels upon diving into the crystal clear underwater world gives rise to the romantic "Ahh," an inner exclamation at penetrating a secret world, an expression of fulfillment of that desire to truly see the Other, the unseeable. Park and garden landscapes were also designed with these moments of surprise at an unexpected sight in mind. Following stretches of wooded area, sudden clearings provided new, unanticipated vistas. For hunters stalking their prey, too, a large part of the appeal is based in the fact that they do not know which individual animals they are hunting and what wildlife they might see. It is the opacity of the forest that makes the hunt magical. It has always been about the sudden change in atmosphere, which is connected to the change in outlook. A fixed nature is no longer enchanting. The magical tension that arises from the unknown is supplanted by new values: predictability and transparency, clarity and structure. This shifts the emphasis from subjective experience to objective understanding. Feelings about nature are replaced by the informed examination thereof, although even this is not wholly emotionless.

Furthermore, the experience of nature will no longer be limited to the time actually spent in nature. The Animal Internet connects us to nature around the clock and from any place on Earth. Whenever we want, we can check and see how our animal friends are faring in the wild. Upon arrival in nature, we will perceive our surroundings in a heightened sensory state, in which virtual and actual reality continuously overlap. This gives rise to an image of nature that can be described as "augmented reality." Augmented reality means seeing more than one truly sees.

This will also help us gradually learn about the relationship between nature and technology differently from how we are used to doing, and from how we have been taught. As paired opposites, nature and technology have shaped the discourse on sustainability in postindustrial society. The fundamental opposition of organic and inorganic systems has become a basic assumption that is no longer rendered problematic, and that therefore stands in the way of new intellectual approaches. One could argue that solar and wind power, as well as geothermal energy use, overcome this dichotomy of nature and technology, or that they are even symbiotic forms of the natural and artifact. The revolution in green energy, however, is itself a good example of the intransigence of this contradiction. After all, we refer to it as *environmental* technology, and not *natural* technology. We thus distinguish between a "nature" that may not come into contact with technology, lest it be contaminated, and an "environment" that is primed for a symbiosis with technical structures, and that is ultimately no longer nature, but a postindustrial space that systematically exploits that which was once "nature." A solar farm is no longer nature; instead, it is an industrial energy field. The same could be said for offshore wind farms. Discourse on sustainability thus intensifies the opposition of nature and technology, rather than resolving it. Animals and technology are kept even more radically apart. We join on the side of the animals in order to count as the good guys, as those who haven't yet given up on our pact with Mother Earth. This overlooks the fact that the survival of many animals today is already dependent on technical structures like satellite imaging and sensors. In short: the image of a technology-free nature is a myth created by humans who, feeling beset by technology, are looking for a way to soothe their souls.

In eliminating the contradiction between nature and technology, a new notion of beauty emerges. Industrial technology—that is, technology that followed manual trades—cannot be beautiful, because its form follows its function, and its design, the laws of serial production. Beauty is individual; the economic logic that underlies technological products defies individuality. Cars, clocks, and toasters are, of course, often marketed using language to suggest they were products of nature. Whenever this happens, though—every time technology is pitched using arguments of aesthetics—an imitation of nature is taking place. Examples of bionic design, such as car fenders or aircraft fuselages, are considered beautiful because we recognize natural forms in them. Technology has not introduced its own beauty. Nature after nature will accomplish this, though. It will establish a beauty based in closeness and authenticity, a beauty that cannot be seen, but that can be experienced only in existential nearness: a beauty of being.

The quality of the images we have of nature will also change. The images will no longer be presented in the high-definition, super slow-motion style that constitutes our current vision of the natural world; instead, they will be unfocussed, pixelated, and blurry snapshots in black and white that come to us from woods, rivers, mountains, and caves. Animal movies of the future will no longer show rare footage of a reclusive panda bear in a remote Chinese province, footage that was the result of weeks-long persistence by dedicated naturalists; instead, they will be seemingly banal images of a fox or a deer in the nearby woods or meadow. These different pictures do not elicit boredom, however; rather, they will create a new authenticity of natural awareness. They will transform the everyday into the exotic.

As a result, we can no longer speak about nature as if it were an alien system. Only then will we experience nature as something that always surrounds us and with which we interact, without even noticing. Our "environment," as we have functionally and pragmatically learned to call it, will regain contours and color. "Nature" is "environment"—it is humans' environment. Nature will thus be experienced as a real, existent ambiance once more, as a pulsating space that envelops humans, and to which humans belong. It will appear as a net of visible and invisible ties that connect all living creatures to one another. Nature cannot be locked out; indeed, it typically disregards the constraints set in place by ecologically active humans. It often returns, unauthorized, to the zones of civilization from which it has been banned. Wild animals that win back urban areas are as visible an expression of this natural force as the plants that shoot out of concrete wastelands. Military training grounds often boast greater biodiversity than some nature preserves, in which decades of money and care have been invested.

Given the extent to which the borders between "nature" and "civilization" are flooded and compromised by nature itself, a new notion of nature emerges. Nature is no longer viewed according to a Rousseauian a priori model, as a situation of total virginity, as an "uncivilization" that may not be entered; instead, the wilderness appears when equilibrium has been established between the powers of nature and civilization. Wilderness need not necessarily be prehuman and isolated, as the Transcendentalists, the fathers of ecological thought, envisioned it. Instead, there is a new form of wilderness through which virtual fences and radio signals course, in which humans move freely and pursue the tasks of civilization, and to which animals adapt by interacting with humans.

The discourse on nature will thereby change its values. It will no longer revolve around the concept of sustainability, which has thus far dominated the leading schools of thought, but will instead concentrate on the idea of resilience. Creating something sustainable means making it keep, removing it from time, preventing it from changing. Panda habitats and untouched rainforests are ideological attempts at conservation. In reality, though, acid rain falls there, too. In reality, contagious animal diseases spread there, too. In reality, invasive species proliferate and destroy native plant and animal populations there, too.

Sustainability's principle of stasis negatively manifests itself in other areas, as well. It was the force behind the construction of LEED-certified, energy efficient skyscrapers in Manhattan. When Hurricane Sandy hit the Eastern seaboard of the United States in 2012, these buildings were the first to flood and lose power. New York would have suffered less damage from Sandy had a lighter, more flexible infrastructure been constructed, with built-in redundancies and energy systems that switch from electricity to solar power or to a gas tank in the basement, in cases of emergency. Buildings such as these would be able to withstand natural disasters, even another flood of the century. They would be intelligent buildings with the ability to repair themselves.

In nature after nature, the objective is to complete the transition from the "ecoperfect" to the "extremophilic"—in other words, a transition from the idea of nature as delicate and fixed to nature as hardy and constantly shifting. It's about establishing a culture of insecurity. Nature preserves and energy-efficient buildings without redundant systems on the one hand, insular national data protection schemes on the other hand are expressions of a culture of security. They are forms of the illusion of control. Resilience, after all, has a psychological dimension, too. We must learn to tolerate contradiction once more, not to dismiss unclear data and vexing notions, and to view fear as just another part of the system. Intolerance of contradiction is replaced by an active tolerance of ambivalence.

A resilient natural and animal world is also dominated by the idea of adaptation and interaction with the environment. The animals of the future will be able to anticipate situations that might endanger them. They can analyze their own status and that of their environment. And they can repair themselves, so to speak, in that they can send humans alerts. The natural system that emerges from this interconnection of human and animal societies is also resilient. By tapping into the animal sensorium, it will become easier for humans to predict and react to extreme events. Animal knowledge enables greater mobility, which itself enables us to better prepare for catastrophes. An ecology of resilience thus emerges within nature after nature.

ECOLOGY AFTER ECOLOGY

"Dark ecology": the state of current thought about nature is worrying. The term "dark ecology" describes the helplessness with which ecology operates, taking aim at its ineffectiveness in regards to the speed at which nature is crumbling in its caring hands. Although conservation efforts have been successful worldwide in realizing one of their central goals—namely, establishing and expanding parks, reserves, and sanctuaries—they are still losing the battle against extinction. In 1950, there were ten thousand nature preserves, and today there are over one hundred thousand areas where human development is restricted. An estimated 13 percent of the earth's surface is protected, which totals an area greater than the size of South America. Nevertheless, the disappearance of species is unstoppable. This shows that sustainability is the wrong strategy.

Furthermore, the assumptions on which the ecology of sustainability is based are often false. Studies have shown that ecosystems recover much more quickly from extreme encroachment and natural disasters than we generally think. Of two hundred and forty habitats that were subjected to destructive human interventions such as deforestation, mining, oil spills, and other pollution, 173 had recovered after five years, both in terms of biodiversity and other quantitative ecological variables.

Faced with these realities, ecological thought in the Anthropocene must reinvent itself. It will naturally still be important to protect biodiverse areas, but this will be only a part of the ecological concept. The bigger questions are: How can humans reestablish contact with nature? How can a new dialogue form between them and animals? Moreover: What will happen to the parts of the planet that are not protected? What will happen to the meadows, forests, rivers, and lakes, in or around which human development must continue? What will happen in the areas in which the primary concern is producing food for starving people? Habitats and nature preserves can be afforded only by those who have enough to eat, or who have enough tax money at their disposal to subsidize domestic agriculture. In places where humans are starving or cannot see a future for themselves, however, this form of barricading and preserving a Western image of "wilderness" is immoral. In a world where over 2.5 billion people live on less than two dollars a day, and where one billion suffer from chronic malnourishment, it's an untenable ambition to restrict human development in large areas and to protect forests from logging.

The ecology of resilience will promote a vision of our planet in which humans and animals coexist—emotionally as well as economically. Technology plays a central role in this newly developed ecology "after nature." Using the notion of a resilient nature as its basis, the aim must be to implement "development by design." Natural spaces must undergo technological development targeted in such a way that humans can live and work within them. There would be necessary compromises to be made, but the majority of technological resources should be introduced to protect the living environments and species serving the largest possible number of people. Instead of reconstructing prehuman landscapes, an ecology of resilience must be measured against its success in revitalizing awareness among city dwellers of the importance of nature and animals. Only then will it fulfill its critical goal of making the usefulness and beauty of nature approachable for humans, thereby motivating them to become cocreators of this beauty.

The ecology of resilience is an ecology without fences. In our attempt to protect nature against technology and civilization, we are constructing a nature that has never actually existed. Nature preserves should signify sustainable counterpoints to the destructive power of progress. They maintain the illusion that a good life can exist within the bad. In reality, however, nature preserves are green ghettos that represent a positive foil to the gray ghettos of the city. Access to both is restricted: to the gray ghetto, for the safety of those entering, and to the green ghetto, for the safety of those living there. In essence, though, both ghettos serve the same purpose: they brush the problems aside. They marginalize. They segregate. The ecology of segregation mirrors the sociology of segregation.

By contrast, the ecology of resilience is an ecology of inclusion. This applies especially to the relationship between humans and animals. Animals that have been reintroduced to the wild require human care and intensive supervision in order to survive in heterogeneous landscapes that may prove dangerous to them. Wild animals must be managed intelligently, in order to find the last green bridges through civilization. Those humans who have grown estranged from nature also require attention, because they are more than just the object of moral admonition and ecological appeals. The primary target group of ecology after ecology is, paradoxically, not animals, but humans.

Nature conservation, which has grown into dogma, is striking in its inherent glut of regulations and restrictions. "Sustainability," "ecology," and "diversity" are fighting words utterly lacking in real substance. They are the semantic holy cows of an over-civilized society, and they construct the image of an abstract, unreal, breakable nature that appears as a strange, seemingly sickly mixture of romantic ideals and moral postulates.

An absurd model reigns supreme, namely that we are saving the earth from humans and practicing a form of conservation that punishes people who are interested in getting closer to nature. The apocalyptic argumentation patterns followed by conservationists, who are forever wagging their fingers and warning of the impending downfall, share one central objective: disseminating guilty consciences.

Nature must return to being something that is not defined first and foremost by restrictions, but that comes from the inside. Humans must be reeducated to view their entering and getting closer to nature not as an anarchistic statement, but as an existential action. We have to learn to feel nature again. Particularly in this regard, adherence to the paradigm of nature preserves is fatal. In systematically cordoning off nature from society, these reserves discriminate against animals by widening the gap separating "us" from "them."

The ecology of resilience, on the other hand, is a social ecology. People who watch animal parenting habits caught on film by cameras installed in incubators, aeries, or otter dens, will develop much stronger emotional ties to wild animals than those exposed even to conservationists' most urgent appeals and arguments. Zoologist Josef Reichholf emphasizes the point that in addressing the public, in particular, new avenues must be explored, and the concepts introduced by wildlife photographer and filmmaker Heinz Sielmann and conservationist Bernhard Grzimek, who were ahead of their time, must be rehabilitated and adapted to contemporary society: "The pioneers of wildlife research in Africa attempted to share the dialogue they had opened with animals by means of books and films. We are still outraged by the big game hunters who unscrupulously shot the tagged lion that had been moving freely through the researchers' camp, threatening no one. But years passed between the time of the event and its reporting. The hunter could no longer be confronted with the public outcry. It's different when the event is experienced in real time by a wide audience—like accompanying the waldrapp flock in the ultralight aircraft from the base in Bavaria or Austria to their Italian wintering grounds, straight over the Alps and the bird hunters' shotguns. The scientist flying along thus becomes the direct conduit for dialogue between thousands of people and the glossy black birds whose survival is at stake. Their needs and hardships become visible, and therefore comprehensible. The same can be said for their ability to communicate with humans."

It would be insufficient if this new nearness to animals served merely as entertainment. It's not enough for viewers to delight in the images of the birds' majestic flight, and to admire the courage of their escort in the aircraft. Social ecology will fundamentally change the dialogue between humans and animals. In the future, humans will be more readily willing to address issues facing animals, or to donate money to combat these issues. When the fear of an animal's defeat translates into a tweet, then the transition into social ecology has been achieved.

Ecology after ecology will see many innovative pedagogical concepts developing, to spark new communication with nature. New disciplines will emerge, like eco-caching, the act of locating animals in the user's vicinity. National parks can launch digital mobile museums featuring a selection of tagged animals representative of the species living there, thus connecting their visitors with the animals. Lifeless museum collections and shelves of preserved specimens and bones, which present nature as something vanquished by civilization, will be replaced by living displays. Active animal tracking will ensue. By developing the appropriate apps, vacation destinations can promote ecotourism, earmarking the resulting revenue for conservation programs. Natural museums will thus become biodomes, in which real time data in visitor-accessible exhibition rooms provide an intimate experience of nature. Once these digital interfaces and spaces exist, animals will again integrate themselves into the world from which we have chased them.

The effects of neogreen ecology can be condensed into the four vertices of knowledge, responsibility, relationship, and communication. Knowledge engenders responsibility, which brings about a relationship, which can then become communication. Knowledge arises from the technological infrastructure of the Animal Internet; responsibility emerges from social ecology; relationships are continually formed on the Animal Internet's social media channels; and communication between the human and animal worlds is the result of this interaction.

There are many reasons to believe in the logic of a postecological discourse, because the situation in which we currently find ourselves—rapidly diminishing biodiversity, ecological systems' continual decline, humans' extensive alienation from nature, nonexistence of a living connection to the animal world—is the result of orthodox ecological thought, the foundations of which were laid in the 1970s and dominate the conversation to this day. Such conservation philosophy condemns nature to disappear from the lives of humans and society, rendering it irrelevant, which pulls the rug out from under the feet of ecological thought and its self-justification. New ecological thought must operate with the concept of natural resilience. It must trust nature to connect with humans and regenerate. It must regard the influence of technology on nature as fundamentally valuable. It must allow a certain degree of economic thought in ecological debate. It must open nature preserves and grant humans access to animals again. Postecological thought may not be allowed to repeat the central mistake made by orthodox ecology, namely the a priori degradation of the animal world to a world of objects that can be exploited for human interests. It must start from scratch and ask itself: *What do animals mean to me?* A new and viable school of thought can develop from that starting point alone.

THE MONKEY-ROPE AND MOBY DICK 2.0

Every version of ecology engenders the culture and artistic creation that it deserves. Viewed in this way, we are not doing so well these days. The nineteenth and early twentieth centuries produced brilliant animal literature. Thoreau, Melville, London, Hemingway, and Kipling, Daudet, Hugo, and Maupassant—they all had an immediate take on nature and animals. The same could be said of Friedrich Schiller's *The Cranes of Ibycus* and Heinrich Heine's *Atta Troll*. In today's literature, by contrast, animals don't play as significant a role. The imaging process has petered out, because the source of these images is no longer accessible.

Biophilia—and this is something that bears repeating—is not only an evolutionarily defined attitude toward nature, but also a powerful engine of culture and the generation of images. Edward O. Wilson demonstrates this implicitly in choosing the image of the snake, which is represented in the same basic way in countless cultures worldwide, to develop his thesis on the ambivalence of biophilia.

Snakes have been feared, and they have been worshiped as gods. Many people feel both drawn to and repelled by snakes. Snakes live in our dreams. They populate the imagination of city dwellers who live at a far remove from nature and have never seen a snake in the wild.

Freud contended that images of snakes were sexual symbols, products of the unconscious that issue from our dreams. Wilson takes a different approach. For him, the persistence and omnipresence of snake imagery can be viewed in a phylogenetic context. He explains the cultural ubiquity of the snake by emphasizing the special role the animal played in human evolution. The human mind does not have the capacity to conceive of reality, in its chaos, as a unified whole. It therefore encrypts critical situations and dangers in images that can be linked back to concrete, empirical experiences. These images are passed along and thus develop into cultural genetic material that serves the survival of the species. Images emerged that functioned as biases toward human surroundings, and the synthesis of all these biases is the basis of human nature. These biases, as Wilson calls them, were necessary for human survival in earlier ages. Frequent early encounters with snakes have led to our general aversion to their physical presence and even images of them, despite their not posing any acute danger to the large majority of humans.

This biophilic dynamic in the generation of images doesn't apply only to snakes, of course, but to other animals and natural phenomena, as well. Images of pigs and rats are further examples of an evolutionarily encoded process of imagination. The process is always the same: observations give rise to a glut of images that do not all, nor always or everywhere, serve the cause of survival. Cultural artifacts are the result of processing these images. They are the result of processing the images that our consciousness has gathered over the course of our evolutionary history of battling nature. Observations and information that humans gather in nature are an important source of human culture. Human consciousness, in interacting with nature, functions as an image generator ceaselessly re-creating the outside world and organizing, arranging, and shaping it by means of symbols, stories, maps. The evolutionary purpose of art is, of course, to keep the memory of human origins alive, thereby maintaining humans' positive relationship with nature. This image generator has stopped running. The Animal Internet can start it up again.

The images that emerge from nature are characterized by the tension between biophilia and biophobia that constitutes the existential dualism between humans and animals. A central scene in Herman Melville's *Moby Dick* illustrates this. One can hold forth endlessly on the big ideas about the connection of all living things, while brilliant writers manage to capture it by means of a single pithy image. Herman Melville did just that; it occurs in chapter seventy-two of his novel. Here, the narrator describes how Ishmael and Queequeg go about flensing the slain whale, the arduous process of stripping its blubber. It is dangerous work. Queequeg stands atop the whale's slippery carcass, which is still in the water. Ishmael, his bowsman, secures him from the deck by means of a rope, to keep him from tumbling into the ocean and falling prey to the many sharks swarming the carcass. This rope has a special name, which provides the title for chapter seventy-two: The Monkey-Rope.

Monkey-ropes are, for one, the vines monkeys use to swing through the jungle. In this simple scene, Melville creates a universal metaphor for human existence within the living system of nature:

> Being the savage's bowsman, that is, the person who pulled the bow-oar in his boat (the second one from forward), it was my cheerful duty to attend upon him while taking that hard-scrabble scramble upon the dead whale's back. You have seen Italian organ-boys holding a dancing-ape by a long cord. Just so, from the ship's steep side, did I hold Queequeg down there in the sea, by what is technically called in the fishery a monkey-rope, attached to a strong strip of canvas belted round his waist.
>
> It was a humorously perilous business for both of us. For, before we proceed further, it must be said that the monkey-rope was fast at both ends; fast to Queequeg's broad canvas belt, and fast to my narrow leather one. So that for better or for worse, we two, for the time, were wedded; and should poor Queequeg sink to rise no more, then both usage and honour demanded, that instead of cutting the cord, it should drag me down in his wake. So, then, an elongated Siamese ligature united us.

"So, then, an elongated Siamese ligature united us"—one would be hard-pressed to express the symbiosis of all living beings in a more succinct, beautiful way. The whale's cleft back becomes a metaphor for the half-liquid, half-solid, and thus swaying world that nourishes humans, but that they must also share with other predators. The monkey-rope ties the one working in nature, that half-wild Queequeg—who has himself only recently emerged from nature and can still communicate with its primal forces— to his observer, the narrator we call Ishmael and who we are, of course, ourselves. The image of the monkey-rope refers to both: as a liana, on the one hand, it evokes the mythical potential of the jungle, the interweaving of all living things, and the natural world's ability to regenerate; as a leash tethering the organ-grinder's pet monkey, on the other hand, it represents the domestication and belittlement, indeed the exhibition of nature. Life, death, destruction, perversion—all semantic levels of nature are linguistically bundled and condensed here. At the heart of this image is the totality of the living environment, in which humans are eternally and inextricably connected to the simultaneously destructive and productive forces of nature.

The best key to interpreting this scene can therefore not be provided by a literary scholar, but must come from Edward O. Wilson. Biophilia, which he defines as "the urge to affiliate with other forms of life," gets to the heart of Melville's powerful scene. For today's readers, the scene comes across as exotic and remote. Moreover, it is superimposed with the morally- and media-driven discussion on the pros and cons of whaling. But deep inside, according to Wilson, our biophilic self is reverberating with a memory. After all, the impetus driving us outdoors today is none other than that which guided our hunter-gatherer ancestors out into the savanna. Millions of years lie between a Sunday stroll in Central Park and a hunt for the now extinct mammoth, but at the same time, no more than a few seconds have passed. Wilson states that we will forever remain alert and alive in the world's vanished forests.

This is why we like to go to the zoo and to watch animal documentaries and cat videos. Anyone who is reflective will discover a daily multitude of other biophilic impulses. The smell of water, the buzzing of a bee, a trampled tuft of grass were observations necessary for our ancestors' survival. Back then, humans *needed* to love nature, in order to survive it. This instinct has not left us. The savanna endures in those who spend their days pounding the pavement, riding the subway trains, and working in the skyscrapers of Tokyo, St. Petersburg, and San Francisco. We remain hunter-gatherers in a world of computers and touch screens.

ACKNOWLEDGEMENTS

The initial idea for this book arose through conversations with Dr. Alexander Kissler, cultural editor at the German monthly magazine *Cicero*. The publication also carried my first report on this topic.

Thanks are due Andreas Rötzer and Tilman Vogt at Matthes & Seitz Berlin for their patience and trust, as well as for their careful work with the manuscript. Thanks to everyone else at the press for their tireless efforts.

Over the course of my research, there were many researchers and scientists who provided me with invaluable references and patiently answered my questions. In first place are Dr. Martin Wikelski of the Max Planck Institute and his team, without whom this book would never have been written. Johannes Fritz of the European Waldrappteam introduced me to the world of the waldrapp. Heartfelt thanks are also due to Markus Bathen (NABU), Dr. Roland Kays (North Carolina State University), Dr. Aline Kühl-Stenzel (UNEP/CMS), Dr. Carola Otterstedt (Bündnis Mensch & Tier), and Dr. Josef Reichholf. Thank you, too, to Dr. Alfred Bach (Heidelberg) and Barbara von Wulffen (Munich) for their motivation and support.

NOTES

2   *Since downloading* Animal Tracker *onto her phone*: This free app can be downloaded from Google Play, iTunes, and the homepage of the Max Planck Institute of Ornithology. Users can participate in various observation projects: www.orn.mpg.de/animaltracker. Cf. Chelsea Wald's report *Follow That Bird: Real-Time Data on Migrating Birds, Coming to a Phone Near You*, at www.earthtouch-news.com/wildlife/conservation/follow-that-bird-real-time-data-on-migrating-birds-coming-to-a-phone-near-you.

    Nature webcams, of which there are now many, provide insight into the lives of animals. Lothar Lenz's YouTube channel, which documents animal life in the Eifel region of Germany, is very impressive: www.youtube.com/channel/UCgJKztDgGQtFu4M_jSEpjcg.

12   *Bit by bit, a transparent natural world will emerge*: The Zoological Society of London has set up small automatic cameras all over the world. The pictures they capture are sent directly to an app on users' smartphones. Users can help researchers identify the photographed animals by matching them to images in the online Field Guide: www.edgeofexistence.org/instantwild.

19   *He describes the bird as*: Cited in Albus, Anita. *On Rare Birds*. Trans. Gerald Chapple. Guildford, Ct: Lyons Press, 2011. Print.

25   *The boys' driving curiosity*: Friedrich Georg Jünger, *Grüne Zweige. Ein Erinnerungsbuch*, Stuttgart: Klett-Cotta, 1978, p. 113-114.

26   *In the foreword of his wonderful book*: Josef H. Reichholf, *Rabenschwarze Intelligenz. Was wir von Krähen lernen können*, Munich: Piper, 2012, p. 7-8.

28   *This faculty the modern age calls common-sense reasoning*: Arendt, Hannah. *The Human Condition*. 2nd ed. Chicago: University of Chicago Press, 1998. Print. p. 284.

30   *There is even a condition*: Richard Louv, *Das Prinzip Natur. Grünes Leben im digitalen Zeitalter*, Weinheim and Basel: Beltz, 2012.

     Far from an alarmist horror scenario: Richard Louv, *Das letzte Kind im Wald*, Weinheim und Basel 2011.

31   *The numbers have decreased by 7 percent*: http://www.fs.fed.us/nrs/pubs/jrnl/2014/nrs_2014_stevens_001.pdf

     This decline, coincident with: http://www.nps.gov/tourism/ResearchTrendsandDatainfo/youthalternativestoparkandoutdoorrecreation.pdf

36   *They have little to no contact*: John Berger, *Why Look at Animals?* London: Penguin, 2009, p. 24: "In the past, families of all classes kept domestic animals because they served a useful purpose—guard dogs, hunting dogs, mice-killing cats, and so on. The practice of keeping animals regardless of their usefulness, the keeping, exactly, of *pets* (in the 16th century the word usually referred to a lamb

raised by hand) is a modern innovation, and, on the social scale on which it exists today, is unique. It is part of that universal but personal withdrawal into the private small family unit, decorated or furnished with mementoes from the outside world, which is such a distinguishing feature of consumer societies."

37   *It ultimately amounts to the same*: Aristocratic practice was the precursor to house pet taxidermy: Condé, Frederick the Great's favorite horse, was prepared and put on display, as was Cosa Rara, the favorite horse of King Ludwig II of Bavaria, which can still be viewed at the Marstallmuseum at Nymphenburg Palace in Munich.

38   *According to the World Wildlife Fund's* Living Planet Index: http://wwf.panda.org/about_our_earth/all_publications/living_planet_report/

       It is now poised to lose:
       http://www.theguardian.com/environment/2014/sep/09/north-america-birds-extinction-study-climate-change

39   *In order to get a feel*: Edward O. Wilson, Bert Hölldobler, *The Ants*, Cambridge, Mass.: Harvard University Press, 1991.

       Stephen Keller, Edward O. Wilson (Eds.), *The Biophilia Hypothesis*, Washington D. C.: Island Press, 1995, p. 36.

40   *The earth is becoming a planet of dying apes:* Cf. "Planet of the dying apes: experts sound alarm over shrinking habitats," http://www.rappler.com/science-nature/environment/61632-experts-sound-alarm-shrinking-habitats.

For example, there are far too few tigers: See Stuart H.M. Butchart et al., "Global Biodiversity: Indicators of Recent Decline," in *Science* 328/2010, p. 1164–1168.

41 *Between 2000 and 2011*: ibid.

43 *Take, for instance, the Spotted-tail Quoll*: Cf. www.theage. com.au/environment/animals/141-years-on-rarest-of-creatures-enters-the-frame-20131002-2usxv.html.

46 *Since the birth of the Internet of Things*: See Peter Paul Verbeek's stance on the Internet of Things in *Moralizing Technology. Understanding and Designing the Morality of Things*, Chicago: University of Chicago Press, 2012.

What about when intelligent technology: Neil Gershenfeld provides an overview in *Wenn die Dinge denken lernen*, Berlin: Econ, 1999.

51 *This technique is frequently used*: See Roland Knauer's "Auf einer Wellenlänge," in *Die Welt,* January 12, 2014, p. 22.

This applies primarily to great white: Cf. www.welt.de/reise/Fern/article123655643/Wie-twitternde-Haie-Wassersportler-schuetzen.html.

52 *Could it actually have been*: A short documentary produced by the Smithsonian Institute illustrates the scene: http://www.youtube.com/watch?v=Z_QyGANCUJI.

53 *This would allow the animals'*: Cf. Jörn auf dem Kampe, "Die Signale der Tiere," in *GEO* 7/2014, p. 119.

56 *Take, for example, the Magellan penguins*: Jan Hoekstra's

essay, "Networking Nature," in *Foreign Affairs*, March/April 2014, p. 82.

59   *The aggregate wisdom*: My thanks to Dr. Alfred Bach of Heidelberg for coining the term *phenome* over the course of our many discussions.

60   *This is the terminus*: Berger, *Why Look at Animals?*, p. 70.

As the English essayist: *Ibid.*, p. 12.

Life without animals: Read a full historical account of human-animal relations in Robert Delort, *Les animaux ont une histoire*, Paris: Seuil, 1984.

62   *The number of animals in France doubled*: According to Éric Baratay, *Bêtes de somme. Des animaux au service des hommes*, Paris: Points Histoires, 2008, p. 11-12.

65   *The L'Aquila earthquake allowed*: Stephen Faris, "Can Toads Predict Earthquakes?" in *Time,* April 1, 2010, http://content.time.com/time/health/article/0,8599,1977090,00.html.

Pilots have observed: Example borrowed from Jacques Perrin's film *Nomaden der Lüfte – Das Geheimnis der Zugvögel*, France etc. 2011.

Animals often communicate: Zoologist Karl von Frisch used the waggle dance of the honey bee to demonstrate how information like the location, quality, or quantity of resources is communicated. Migrations can be prompted by threats to the animals or food shortages, for instance.

67   *The furry bats are cooked whole:* Thomas Scheen, "Flughunde dürft ihr jetzt nicht mehr essen," *Frankfurter Allgemeine Zeitung* March 27, 2014, p. 7, available online at: http://www.faz.net/aktuell/gesellschaft/gesundheit/ebola-ausbruch-in-guinea-das-virus-breitet-sich-rasch-aus-12865408.html.

68   *The fruit bats don't let that stop them*: See Catarina Pietschmann, "Bewegtes Leben," in *Max-Planck Forschung* 2/2012, p. 26–33, available online at: www.mpg.de/6633612/tierwanderungen.

70   *The new strategy to save the world's fauna*: Edward O. Wilson, *The Future of Life*, New York: Random House, 2002. p. 151.

73   *Using this data, whale sharks*: Casey Cazan, "Hubble Space Telescope Technology Links with Internet to Save Endangered Species," http://www.dailygalaxy.com/my_weblog/2007/09/httpwwwitworldc.html.

74   *At the same time, modern-day naturalists*: See Ottmar Ette, *Alexander von Humboldt und die Globalisierung: Das Mobile des Wissens*, Frankfurt am Main 2009.

75   *I grew up in a small farming town*: All quotations here and following drawn from personal conversations between the author and Professor Martin Wikelski.

83   *These birds spend the spring:* Cf. the study on Swainson's hawks at: www.movebank.org/node/2064.

84   *Since rangers and conservationists*: Images available here:

http://www.treehugger.com/endangered-species/worlds-rarest-gorilla-caught-film-cameroon.html.

The result is systematic data: Richard Bergl, Andrew Dunn, Aaron Nicholas, "Using technology and Partnerships to Save the Critically Endangered Cross River Gorilla," www.aza.org/cross-river-gorilla.

85   *Nature has always been man-dominated*: Gayathri Vaidyanathan, "Gorilla Populations Need More Human Interference,"http://discovermagazine.com/2013/may/06-tracking-gorilla-populations-human-interference.

87   *Finally, knowing the animals' movement patterns*: See Steffan Zuther, "Durchs weite Kasachstan," in *ZGF Gorilla*, 3/2010, p. 11–13.

In Southern Nepal, a GPS device: "Nepal Uses Satellites to Track Rare Tiger," available online at: www.mnn.com/earth-matters/animals/stories/nepal-uses-satellites-to-track-rare-tiger.

88   *Elephants are so intelligent*: Jan Hoekstra, "Networking Nature," pp. 80-82.

89   *That's the lesson we've learned:* Personal conversation with the author. Farming apps already exist that use chip technology to help monitor the health of cows in large herds. (www.thingworx.com/2014/04/vital-herd-selects-thingworx-m2m-technology-platform-for-livestock-management) Several farmers have even started allowing their cows to tweet about their living conditions and milk production, in order to give consumers a behind-the-scenes

look at life in the barn (http://mashable.com/2010/04/27/cows-on-twitter).

91 *It doesn't just apply to the lion*: Personal conversation with the author.

92 *Knowledge of wolves is minimal*: Personal conversation with the author.

113 *"We should not," says zoologist Reichholf in this regard*: Personal conversation with the author.

117 *To do such a thing would amount to a breach of protocol*: Richard Nelson, "Searching for the Lost Arrow. Physical and Spiritual Ecology in the Hunter's World," in Stephen R. Kellert, Edward O. Wilson, *The Biophilia Hypothesis*, Washington: Island Press, 1993, p. 213.

118 *A quick look back at the history of philosophy*: See Melchior Westhues, *Über den Schmerz der Tiere*, Munich: M. Hueber, 1955, p. 6-7.

119 *In Critique of Judgment, Immanuel Kant*: Kant, Immanuel. *Critique of Judgment*. Trans. Werner S. Pluhar. Indianapolis/Cambridge: Hackett Publishing Company, 1987.

According to Schopenhauer: See Oliver Hallich, *Mitleid und Moral. Schopenhauers Leidensethik und die moderne Moralphilosophie*, Würzburg 1998, as well as Rainer E. Wiedenmann, *Tiere, Moral und Gesellschaft. Elemente und Ebenen humanistischer Sozialität*, Wiesbaden 2009.

120 *Close cohabitation with animals*: See David Berreby's lucid

study in *Us and Them. The Science of Identity*, Chicago: University of Chicago Press, 2008.

121 *Humans lived in unity with their natural surroundings*: In *Totemism*, Claude Lévi-Strauss points out that humans used animal diversity to describe social differentiation within their own society. Gilles Deleuze and Félix Guattari, on the other hand, contend that animals first become animals when they flock together, forming a pack. See *A Thousand Plateaus: Capitalism and Schizophrenia*, Minneapolis: University of Minnesota Press, 1997.

122 *Biotechnology has managed to disassemble animals*: Cf. *Frankenstein's Cat: Cuddling Up to Biotech's Brave New Beasts*, New York: Farrar, Straus & Giroux, 2013.

123 *The French philosopher Elisabeth de Fontenay*: Elisabeth de Fontenay, *Le silence de bêtes: La philosophie à l'épreuve de l'animalité*, Paris: Fayard, 1998.

And with that, "the Animal" loses the very last remnants of agency: Admittedly, this assessment is based in the classic understanding of genetics, in which genes serve "only" as an individual's morphological blueprint. Today's understanding of genetics extends beyond this model. The information passed on in the cytoplasm is now considered more important, but it cannot (yet?) be cloned. This means that an individual animal is going to have much more to offer than its morphologically similar, "cloned" counterpart. For this reason, the practice of cloning is unlikely to continue, because it does not optimize an individual animal's productivity levels in captivity as much as had been hoped. Traditional breeding, which is based in statistics, works much more effectively and efficiently at achieving this end. Most importantly, though, when it comes to house pets, "Fifi's" exact characteristics cannot be resurrected.

124 *They are then reflected back only on themselves*: The only other way to assure oneself of the Other contained within oneself is through the heterotopias described by Michel Foucault, and their inhabitants—fools, prostitutes, criminals. Cf. Michel Foucault, *Die Heterotopien/Der utopische Körper. Zwei Radiovorträge*, Trans. Michael Bischoff, Frankfurt am Main: Suhrkamp Verlag, 2005.

To explain this process, historian Richard W. Bulliet: Richard W. Bulliet, *Hunters, Herders, and Hamburgers: The Past and Future of Human-Animal Relationships*, New York: Columbia University Press, 2005.

125 *Vegans with lots of house pets*: Jonathan Safran Foer, *Eating Animals*, New York: Little, Brown & Co. 2009.

126 *The scene was caught on home video*: See www.youtube. com/watch?v=gBpKgykXXRo.

French philosopher Jacques Derrida: Cf. Derrida, Jacques. "The Animal That Therefore I Am (More to Follow)." Trans. David Wills. *Critical Inquiry* 28.2 (Winter 2002): 369-418. Web, March 19, 2015. http://www.jstor.org/ stable/1344276.

127 *Regardless of whether Binti Jua acted more like a human or an ape*: Cf. Andreas Möller, *Das grüne Gewissen. Wenn die Natur zur Ersatzreligion wird*, Munich: Carl Hanser Verlag, 2013.

This is not an arrangement of man and beast: Deleuze, Gilles. *Francis Bacon: The Logic of Sensation*. Trans. Daniel W. Smith. London: Continuum, 2003. Print.

129 *Cat pictures are members of the genus* memes: Richard Dawkins, *The Selfish Gene*, 2nd Ed., Oxford, England: Oxford University Press, 1989, p. 192.

133 *The French garden uses natural elements*: It is critical to note that the use of artistic measures to reconstruct nature is a genuinely European idea. Japanese culture, for instance, sees nature exclusively as something for art to conquer.

140 *The* New York Times *used this as an occasion*: Virginia Morel, *Hunters Kill Another Radio-Collared Yellowstone National Park Wolf*, http://news.sciencemag.org/people-events/2012/12/hunters-kill-another-radio-collared-yellowstone-national-park-wolf.

On June 8, 2014: Darryl Fears, "A Lonely Wolf Gets a New Mate, Powerful Friends and a Little Protection," in *The Washington Post*, June 8, 2014, http://www.washingtonpost.com/national/health-science/a-lonely-wolf-gets-a-new-mate-powerful-friends-and-a-little-protection/2014/06/08/5c990386-eda8-11e3-b84b-3393a45b80f1_story.html.

For more on Wolf OR-7 visit the California Department of Fish and Wildlife website, http://www.dfg.ca.gov/wildlife/nongame/wolf/OR7story.html, and the article by Emily Anthes, "Tracking the Pack," in *The New York Times*, February 4, 2013, A17, www.nytimes.com/2013/02/04/opinion/tracking-the-pack.html?_r=0.

141　*This wolf's life, which is not yet over, was retold in a short film*: The film and other materials available at: http://www.oregonlive.com/pacific-northwest-news/index.ssf/2014/05/oregon_wolf_or-7_finds_a_mate.html.

142　*Researchers have already generated these images*: So-called "Crittercams" are used to explore the deep sea, and are fastened to the bodies of whales and other marine mammals using suction cups. The result: spectacular images captured from new perspectives: http://animals.nationalgeographic.com/animals/crittercam/?rptregcta=reg_free_np&rptregcampaign=20131016_rw_membership_r1p_intl_ot_w - .

　　Lightweight cameras attached to the heads of falcons and eagles document these birds' amazing hunting techniques: http://www.youtube.com/watch?v=yf-VHrmREuA and www.youanimal.it/sulle-ali-dellaquila-sopra-il-monte-bianco.

The result is an entirely new kind of sensory experience: Many more examples are available on this book's Facebook page: http://www.facebook.com/pages/Internet-der-Tiere-Internet-of-Animals/519370431472423?ref=hl.

Many life stories of actual animals: The stork diaries initiave is documented at: http://www.nabu.de/aktionenundprojekte/weissstorchbesenderung/tagebuch. National Geographic's program can be seen at http://education.nationalgeographic.com/education/media/tracking-animal-migrations/?ar_a=1.

143　*These data are then fed into a text generator:* http://redkite.abdn.ac.uk

144 *This was particularly the case for Tolosa*: Tolosa's story is documented on the Museum of Toulouse website: http://www.museum.toulouse.fr/-/dans-les-yeux-de-l-ourse-via-une-camera-embarquee?redirect=%2Fexplorer.

145 *There is no path leading 'back to nature.'*: Personal conversation with the author.

147 *In* The First Life of Saint Francis: Brother Thomas of Celano, *The Lives of S. Francis of Assisi*. Trans. A. G. Ferrers Howell, LL.M. London: Methuen & Co. n.d. Print. https://archive.org/stream/livesfrancisas00howegoog - page/n102/mode/2up.

They are—to reference Buber once more—an "object": Martin Buber, *I And Thou*. http://archive.org/stream/writingsofmartin007421mbp/writingsofmartin007421mbp_djvu.txt.
This theory is endorsed by Carola Otterstedt of Bündnis Mensch & Tier, who also advocates for animals to be recognized as having personalities and dignity.

148 *He sees an emotional progression*: The website Wildearth hosts a number of wildlife cams, documenting the daily lives of many wild animals: http://www.wildearth.tv.

150 *And that the invention thus divinely honored and distinguished*: Thoreau, Henry David. *The Journal 1837–1861*. Ed. Damion Searls. New York: NYRB Classics, 2009. Print. p. 81.

157 *The Anthropocene is the "age of the human."*: This expression was coined in 2002 by Paul Crutzen, winner of the

1995 Nobel Prize in Chemistry: "Geology of Mankind," in *Nature* Vol 415, January 3, 2002, p. 23, available online at: http://www.studgen.uni-mainz.de/sose04/schwerp3/expose/geology.pdf; see also: Will Steffen, Paul J. Crutzen, John R. McNeill, "The Anthropocene: Are Humans Now Overwhelming the Great Forces of Nature," in *AMBIO. A Journal of the Human Environment* 36/8, 2007, p. 614–621; as well as Elisabeth Kolbert, "Anthropozän – Das Zeitalter des Menschen": http://www.nationalgeographic.de/reportagen/anthropozaen-das-zeitalter-des-menschen.

158 *The Animal Internet is thus a symptom*: See also Christian Schwägerl, *Die analoge Revolution*, Munich: Riemann Verlag, 2014.

163 *Nature is no longer viewed according to a Rousseauian a priori model*: See Emma Marris, *Rambunctious Garden. Nature in a Post-Wild World*, New York: Bloomsbury, 2011; as well as: Timothy Morton, *The Ecological Thought*, Cambridge, Mass.: Harvard University Press, 2010.

Wilderness need not necessarily be prehuman: In his essay "Nature," Ralph Waldo Emerson writes, "To go into solitude, a man needs to retire as much from his chamber as from society." https://archive.org/details/naturemunroe00emerrich

164 *When Hurricane Sandy hit the Eastern seaboard*: Robert Capse, "Bounce Back. Why We'd Be Better Off Adapting to Disaster Than Fighting It," in *WIRED*, January 2013, p. 22.

In nature after nature: See Andrew Zolli: *Resilience: Why Things Bounce Back*, New York: Free Press, 2012.

165 *The term "dark ecology"*: This expression was coined by Paul Kingsnorth, "Dark Ecology. Searching for Truth in a Post-Green World," *ORION Magazine* January/February 2013, available online at: http://www.orionmagazine.org/index.php/articles/article/7277.

166 *An estimated 13 percent of the earth's surface*: Cf. United Nations Environment Programme World Conservation Monitoring Centre world database on protected areas. Available online at: http://protectedplanet.net.

Of two hundred and forty habitats: H. P. Jones, O. J. Schmitz, *Rapid Recovery of Damaged Ecosystems*, available online: http://www.plosone.org/article/info%3Adoi%2F10.1371%2Fjournal.pone.0005653.

167 *In our attempt to protect nature*: See Timothy Morton, *Ecology Without Nature. Rethinking Environmental Aesthetics*, Cambridge, Mass.: Harvard University Press, 2007.

169 *Zoologist Josef Reichholf emphasizes the point*: Personal conversation with the author.

172 *In today's literature, by contrast*: Brigitte Kronauer's *Enten und Knäckebrot* and Marcel Beyers's *Flughunde* are exceptions to the very rule they prove.

173 *It therefore encrypts critical situations*: Wilson, *Biophilia*, p. 83-101.

174 *A central scene in Herman Melville's Moby Dick*: Melville, Herman. *Moby Dick; or, The Whale.* http://www.gutenberg.org/files/2701/2701-h/2701-h.htm - link2HCH0072

BIBLIOGRAPHY

Albus, Anita. *On Rare Birds*. Trans. Gerald Chapple. Guildford, Ct: Lyons Press, 2011.

Anthes, Emily, *Frankenstein's Cat: Cuddling Up to Biotech's Brave New Beasts*, New York: Farrar, Straus & Giroux, 2013.

Arendt, Hannah, *The Human Condition*. 2nd ed. Chicago: University of Chicago Press, 1998.

Baratay, Éric, *Bêtes de somme. Des animaux au service des hommes*, Paris: Points Histoires, 2008.

Berger, John, *Why Look at Animals?*, London: Penguin, 2009.

Berreby, David, *Us and Them: The Science of Identity*, Chicago: University of Chicago Press, 2008.

Buber, Martin, *Ich und Du*, Heidelberg: Schneider, 1997.

Bulliet, Richard, *Hunters, Herders, and Hamburgers: The Past and Future of Human-Animal Relationships*, New York: Columbia University Press, 2005.

Butchart Stuart H. M. et al., "Global Biodiversity: Indicators of Recent Decline," in: *Science* 328/2010.

Crutzen, Paul, "Geology of Mankind," in: *Nature* Vol. 415, January 3, 2002.

Deleuze, Gilles. *Francis Bacon: The Logic of Sensation*. Trans. Daniel W. Smith. London: Continuum, 2003.

Delort, Robert, *Les animaux ont une histoire*, Paris: Seuil, 1984.

Derrida, Jacques. "The Animal That Therefore I Am (More to Follow)." Trans. David Wills. *Critical Inquiry* 28.2 (Winter 2002): 369-418. Web. March 19, 2015. http://www.jstor.org/stable/1344276.

Ette, Ottmar, *Alexander von Humboldt und die Globalisierung: Das Mobile des Wissens*, Frankfurt am Main: Insel Verlag, 2009.

Foer, Jonathan Safran, *Eating Animals*, New York: Little, Brown & Co. 2009.

Fontenay, Elisabeth de, *Le silence de bêtes: La philosophie* à *l'épreuve de l'animalité*, Paris: Fayard, 1998.

Foucault, Michel, *Die Heterotopien/Der utopische Körper. Zwei Radiovorträge*, Trans. Michael Bischoff, Frankfurt am Main: Suhrkamp Verlag, 2005.

Gershenfeld, Neil, *Wenn die Dinge denken lernen*, Berlin: Econ, 1999.

Hallich, Oliver, *Mitleid und Moral. Schopenhauers Leidensethik und die moderne Moralphilosophie*, Königshausen & Neumann, Würzburg 1998.

Herzog, Hal, *Wir streicheln und wir essen sie. Unser paradoxes Verhältnis zu Tieren*, Munich: Carl Hanser Verlag, 2012.

Hoekstra, Jon, "Networking Nature," in: *Foreign Affairs*, March/April 2014.

Jünger, Friedrich Georg, *Grüne Zweige. Ein Erinnerungsbuch*, Stuttgart: Klett-Cotta, 1978.

Keller, Stephan, Edward O. Wilson (editors), *The Biophilia Hypothesis*, Washington D. C.: Island Press, 1995.

Kingsnorth, Paul, "Dark Ecology. Searching for Truth in a Post-Green World," *ORION* Magazine January/February 2013.

Louv, Richard, Das Prinzip Natur. *Grünes Leben im digitalen Zeitalter*, Weinheim and Basel: Beltz, 2012.

Marris, Emma, *Rambunctious Garden. Nature in a Post-Wild World*, New York: Bloomsbury, 2011.

Melville, Herman, *Moby Dick*, New York, Macmillan, 1962.

Möller, Andreas, *Das grüne Gewissen. Wenn die Natur zur Ersatzreligion wird*, Munich: Carl Hanser Verlag, 2013.

Morton, Timothy, *Ecology Without Nature. Rethinking Environmental Aesthetics*, Cambridge, Mass.: Harvard University Press, 2007.

The Ecological Thought, Cambridge, Mass.: Harvard University Press, 2010.

Negroponte, Nicholas, "Beyond Digital," in: WIRED 6.12, December 1998.

Reichholf, Josef H., *Rabenschwarze Intelligenz. Was wir von Krähen lernen können*, Munich: Piper, 2012.

Smith, Cyrill, *A Search for Structure. Selected Essays on Science, Art and History*, Cambridge, Mass.: MIT Press, 1981.

Schwägerl, Christian, *Die analoge Revolution*, Munich: Riemann Verlag, 2014.

Thoreau, Henry David. *The Journal 1837–1861*. Ed. Damion Searls. New York: NYRB Classics, 2009.

Tributsch, Helmut, *Wenn die Schlangen erwachen. Erdbebenforscher lernen von Tieren*, Munich: Deutsche Verlagsanstalt, 1978.

Verbeek, Peter-Paul, *Moralizing Technology. Understanding and De- signing the Morality of Things*, Chicago: University of Chicago Press, 2012.

Westhues, Melchior, *Über den Schmerz der Tiere*, Munich: M. Hueber, 1955.

Wiedenmann, Rainer E., Tiere, *Moral und Gesellschaft. Elemente und Ebenen humanistischer Sozialität*, Wiesbaden: VS Verlag, 2009.

Wilson, Edward O., *Biophilia: The Human Bond with Other Species*, Cambridge, Mass.: Harvard University Press, 1984.

Wilson, Edward O., Bert Hölldobler, *The Ants*, Cambridge, Mass.: Harvard University Press, 1991.

Wilson, Edward O., *The Future of Life*, New York: Knopf, 2002.

Zolli, Andrew, *Resilience: Why Things Bounce Back*, New York: Free Press, 2012.

INDEX

CREDITS FOR PHOTOS OF CREATURES EQUIPPED WITH TRACKING DEVICES

THE 6:41 TO PARIS BY JEAN-PHILIPPE BLONDEL

Cécile, a stylish 47-year-old, has spent the weekend visiting her parents outside Paris. By Monday morning, she's exhausted. These trips back home are stressful and she settles into a train compartment with an empty seat beside her. But it's soon occupied by a man she recognizes as Philippe Leduc, with whom she had a passionate affair that ended in her brutal humiliation 30 years ago. In the fraught hour and a half that ensues, Cécile and Philippe hurtle towards the French capital in a psychological thriller about the pain and promise of past romance.

ON THE RUN WITH MARY BY JONATHAN BARROW

Shining moments of tender beauty punctuate this story of a youth on the run after escaping from an elite English boarding school. At London's Euston Station, the narrator meets a talking dachshund named Mary and together they're off on escapades through posh Mayfair streets and jaunts in a Rolls-Royce. But the youth soon realizes that the seemingly sweet dog is a handful; an alcoholic, nymphomaniac, drug-addicted mess who can't stay out of pubs or off the dance floor. *On the Run with Mary* mirrors the horrors and the joys of the terrible 20th century.

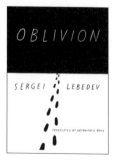

OBLIVION BY SERGEI LEBEDEV

In one of the first 21st century Russian novels to probe the legacy of the Soviet prison camp system, a young man travels to the vast wastelands of the Far North to uncover the truth about a shadowy neighbor who saved his life, and whom he knows only as Grandfather II. Emerging from today's Russia, where the ills of the past are being forcefully erased from public memory, this masterful novel represents an epic literary attempt to rescue history from the brink of oblivion.

THE LAST WEYNFELDT BY MARTIN SUTER

Adrian Weynfeldt is an art expert in an international auction house, a bachelor in his mid-fifties living in a grand Zurich apartment filled with costly paintings and antiques. Always correct and well-mannered, he's given up on love until one night—entirely out of character for him—Weynfeldt decides to take home a ravishing but unaccountable young woman and gets embroiled in an art forgery scheme that threatens his buttoned up existence. This refined page-turner moves behind elegant bourgeois facades into darker recesses of the heart.

THE LAST SUPPER BY KLAUS WIVEL

Alarmed by the oppression of 7.5 million Christians in the Middle East, journalist Klaus Wivel traveled to Iraq, Lebanon, Egypt, and the Palestinian territories to learn about their fate. He found a minority under threat of death and humiliation, desperate in the face of rising Islamic extremism and without hope their situation will improve. An unsettling account of a severely beleaguered religious group living, so it seems, on borrowed time. Wivel asks, Why have we not done more to protect these people?

GUYS LIKE ME BY DOMINIQUE FABRE

Dominique Fabre, born in Paris and a life-long resident of the city, exposes the shadowy, anonymous lives of many who inhabit the French capital. In this quiet, subdued tale, a middle-aged office worker, divorced and alienated from his only son, meets up with two childhood friends who are similarly adrift. He's looking for a second act to his mournful life, seeking the harbor of love and a true connection with his son. Set in palpably real Paris streets that feel miles away from the City of Light, a stirring novel of regret and absence, yet not without a glimmer of hope.

KILLING AUNTIE BY **ANDRZEJ BURSA**

A young university student named Jurek, with no particular ambitions or talents, finds himself with nothing to do. After his doting aunt asks the young man to perform a small chore, he decides to kill her for no good reason other than, perhaps, boredom. This short comedic masterpiece combines elements of Dostoevsky, Sartre, Kafka, and Heller, coming together to produce an unforgettable tale of murder and—just maybe—redemption.

I CALLED HIM NECKTIE BY **MILENA MICHIKO FLAŠAR**

Twenty-year-old Taguchi Hiro has spent the last two years of his life living as a hikikomori—a shut-in who never leaves his room and has no human interaction—in his parents' home in Tokyo. As Hiro tentatively decides to reenter the world, he spends his days observing life from a park bench. Gradually he makes friends with Ohara Tetsu, a salaryman who has lost his job. The two discover in their sadness a common bond. This beautiful novel is moving, unforgettable, and full of surprises.

WHO IS MARTHA? BY **MARJANA GAPONENKO**

In this rollicking novel, 96-year-old ornithologist Luka Levadski foregoes treatment for lung cancer and moves from Ukraine to Vienna to make a grand exit in a luxury suite at the Hotel Imperial. He reflects on his past while indulging in Viennese cakes and savoring music in a gilded concert hall. Levadski was born in 1914, the same year that Martha—the last of the now-extinct passenger pigeons—died. Levadski himself has an acute sense of being the last of a species. This gloriously written tale mixes piquant wit with lofty musings about life, friendship, aging and death.

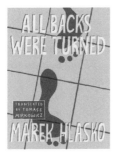

ALL BACKS WERE TURNED BY MAREK HLASKO

Two desperate friends—on the edge of the law—travel to the southern Israeli city of Eilat to find work. There, Dov Ben Dov, the handsome native Israeli with a reputation for causing trouble, and Israel, his sidekick, stay with Ben Dov's younger brother, Little Dov, who has enough trouble of his own. Local toughs are encroaching on Little Dov's business, and he enlists his older brother to drive them away. It doesn't help that a beautiful German widow is rooming next door. A story of passion, deception, violence, and betrayal, conveyed in hard-boiled prose reminiscent of Hammett and Chandler.

ALEXANDRIAN SUMMER BY YITZHAK GORMEZANO GOREN

This is the story of two Jewish families living their frenzied last days in the doomed cosmopolitan social whirl of Alexandria just before fleeing Egypt for Israel in 1951. The conventions of the Egyptian upper-middle class are laid bare in this dazzling novel, which exposes sexual hypocrisies and portrays a vanished polyglot world of horse racing, seaside promenades and nightclubs.

COCAINE BY PITIGRILLI

Paris in the 1920s—dizzy and decadent. Where a young man can make a fortune with his wits ... unless he is led into temptation. Cocaine's dandified hero Tito Arnaudi invents lurid scandals and gruesome deaths, and sells these stories to the newspapers. But his own life becomes even more outrageous when he acquires three demanding mistresses. Elegant, witty and wicked, Pitigrilli's classic novel was first published in Italian in 1921 and retains its venom even today.

***Killing the Second Dog* by Marek Hlasko**
Two down-and-out Polish con men living in Israel in the 1950s scam an American widow visiting the country. Robert, who masterminds the scheme, and Jacob, who acts it out, are tough, desperate men, exiled from their native land and adrift in the hot, nasty underworld of Tel Aviv. Robert arranges for Jacob to run into the widow who has enough trouble with her young son to keep her occupied all day. What follows is a story of romance, deception, cruelty and shame. Hlasko's writing combines brutal realism with smoky, hard-boiled dialogue, in a bleak world where violence is the norm and love is often only an act.

***The Missing Year of Juan Salvatierra* by Pedro Mairal**
At the age of nine, Juan Salvatierra became mute following a horse riding accident. At twenty, he began secretly painting a series of canvases on which he detailed six decades of life in his village on Argentina's frontier with Uruguay. After his death, his sons return to deal with their inheritance: a shed packed with rolls over two miles long. But an essential roll is missing. A search ensues that illuminates links between art and life, with past family secrets casting their shadows on the present.

***Fanny von Arnstein: Daughter of the Enlightenment* by Hilde Spiel**
In 1776 Fanny von Arnstein, the daughter of the Jewish master of the royal mint in Berlin, came to Vienna as an 18-year-old bride. She married a financier to the Austro-Hungarian imperial court, and hosted an ever more splendid salon which attracted luminaries of the day. Spiel's elegantly written and carefully researched biography provides a vivid portrait of a passionate woman who advocated for the rights of Jews, and illuminates a central era in European cultural and social history.

THE GOOD LIFE ELSEWHERE BY VLADIMIR LORCHENKOV

The very funny—and very sad—story of a group of villagers and their tragicomic efforts to emigrate from Europe's most impoverished nation to Italy for work. An Orthodox priest is deserted by his wife for an art-dealing atheist; a mechanic redesigns his tractor for travel by air and sea; and thousands of villagers take to the road on a modern-day religious crusade to make it to the Italian Promised Land. A country where 25 percent of its population works abroad, remittances make up nearly 40 percent of GDP, and alcohol consumption per capita is the world's highest – Moldova surely has its problems. But, as Lorchenkov vividly shows, it's also a country whose residents don't give up easily.

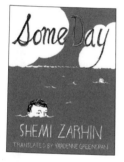

SOME DAY BY SHEMI ZARHIN

On the shores of Israel's Sea of Galilee lies the city of Tiberias, a place bursting with sexuality and longing for love. The air is saturated with smells of cooking and passion. *Some Day* is a gripping family saga, a sensual and emotional feast that plays out over decades. This is an enchanting tale about tragic fates that disrupt families and break our hearts. Zarhin's hypnotic writing renders a painfully delicious vision of individual lives behind Israel's larger national story.

New Vessel Press

To purchase these titles and for more information please visit newvesselpress.com.